CISM COURSES AND LECTURES

The series presents lecture notes, monographs, edited works and proceedings in the field of Mechanics, Engineering, Computer Science and Applied Mathematics.
Purpose of the series is to make known in the international scientific and technical community results obtained in some of the activities organized by CISM, the International Centre for Mechanical Sciences.

INTERNATIONAL CENTRE FOR MECHANICAL SCIENCES

COURSES AND LECTURES - No. 483

NONLINEAR WAVES IN FLUIDS: RECENT ADVANCES AND MODERN APPLICATIONS

EDITED BY

ROGER GRIMSHAW
LOUGHBOROUGH UNIVERSITY, UK

SpringerWien NewYork

The publication of this volume was co-sponsored and co-financed by the UNESCO Venice Office - Regional Bureau for Science in Europe (ROSTE) and its content corresponds to a CISM Advanced Course supported by the same UNESCO Regional Bureau.

This volume contains 31 illustrations

In order to make this volume available as economically and as rapidly as possible the authors' typescripts have been reproduced in their original forms. This method unfortunately has its typographical limitations but it is hoped that they in no way distract the reader.

ISBN 3-211-25259-2 SpringerWienNewYork

PREFACE

Although nonlinear waves occur in nearly all branches of physics and engineering, there is an amazing degree of agreement about the fundamental concepts and the basic paradigms. The underlying unity of the theory for linearized waves is already well-established, with the importance of such universal concepts as group velocity and wave superposition. For nonlinear waves the last few decades have seen the emergence of analogous unifying comcepts. The pervasiveness of the soliton concept is amply demonstrated by the ubiquity of such models as the Korteweg-de Vries equation and the nonlinear Schrodinger equation. Similarly, there is a universality in the study of wave-wave interactions, whether deterministic or statistical, and in the recent developments in the theory of wave-mean flow interactions. The aim of this text is to present the basic paradigms of weakly nonlinear waves in fluids.

This book is the outcome of a CISM Summer School held at Udine from September 20-24, 2004. Like the lectures given there the text covers asymptotic methods for the derivation of canonical evolution equations, such as the Korteweg-de Vries and nonlinear Schrodinger equations, descriptions of the basic solution sets of these evolution equations, and the most relevant and compelling applications. These themes are interlocked, and this will be demonstrated throughout the text . The topics address any fluid flow application, but there is a bias towards geophysical fluid dynamics, reflecting for the most part the areas where many applications have been found.

There are six chapters covering the following themes, with the respective authors named in brackets.

1. Weakly nonlinear long waves and the Korteweg-de Vries equation, with applications to internal solitary waves in the ocean and atmosphere (Roger Grimshaw).
2. The nonlinear Schrodinger equation for weakly nonlinear wave packets with applications to water waves (Frederic Dias and Tom Bridges).
3. Resonant wave interactions, triads and quartets with applications to Rossby and internal waves (Jacques Vanneste).
4. Wave-mean flow interaction for small-amplitude waves, with application to shear flows in the atmosphere and ocean (Oliver Buhler).
5. Weak-wave turbulence and statistical theories, with applications to water waves, Rossby waves and internal waves (Vladimir Zeitlin).
6. A topical example, Bose-Einstein condensates, where the same concepts arise in a different physical concept (Guoxiang Huang).

This text, like the lectures on which it is based, is directed at beginning researchers, postgraduate students, postdoctoral fellows and those researchers wishing to change fields. We hope that it will provide the basic material for the modern theory of nonlinear waves, together with a flavour of the applications in several areas in fluid dynamics.

Roger Grimshaw

CONTENTS

Korteweg de-Vries Equation

Roger Grimshaw[*]

[*] Department of Mathematical Sciences, Loughborough University, UK

Abstract In this chapter we consider weakly nonlinear long waves. Here the basic paradigm is the well-known Korteweg-de Vries equation and its solitary wave solution. We present a brief historical discussion, followed by a typical derivation in the context of internal and surface water waves. Then we describe two extensions, the first to the variable-coefficient Korteweg-de Vries equation for the description of solitary waves in a variable environment, and the second to the forced Korteweg-de Vries equation and the theory of undular bores.

1 Introduction

1.1 Canonical Korteweg-de Vries equation and solitary waves

The Korteweg-de Vries (KdV) equation is

$$A_t + 6AA_x + A_{xxx} = 0, \tag{1.1}$$

written here in canonical form. In (1.1) $A(x,t)$ is an appropriate field variable, t is time, and x is a space coordinate in the direction of propagation. The KdV equation is widely recognised as a paradigm for the description of weakly nonlinear long waves in many branches of physics and engineering. It describes how waves evolve under the competing but comparable effects of weak nonlinearity and weak dispersion. Indeed, if it is supposed that x-derivatives scale as ϵ where ϵ is the small parameter characterising long waves (i.e. typically the ratio of a relevant background length scale to a wavelength scale), then the amplitude scales as ϵ^2 and the time evolution takes place on a scale of ϵ^{-3}.

The KdV equation is characterised by its solitary wave solutions,

$$A = a\,\text{sech}^2(\gamma(x - Vt)) \qquad \text{where} \qquad V = 2a = 4\gamma^2. \tag{1.2}$$

This solution describes a family of steady isolated wave pulses of positive polarity, characterized by the wavenumber γ; note that the speed V is proportional to the wave amplitude a, and to the square of the wavenumber γ^2.

1.2 Brief history

The KdV equation (1.1) owes its name to the famous paper of Korteweg and de Vries, published in 1895, in which they showed that small-amplitude long waves on the free surface of water could be described by the equation

$$\zeta_t + c\zeta_x + \frac{3c}{2h}\zeta\zeta_x + \frac{ch^2}{6}(1 - \frac{W}{3})\zeta_{xxx} = 0. \tag{1.3}$$

Here $\zeta(x,t)$ is the elevation of the free surface relative to the undisturbed depth h, $c = (gh)^{1/2}$ is the linear long wave phase speed, and $W = T/gh^2$ is the Weber number measuring the effects of surface tension (ρT is the coefficient of surface tension and ρ is the water density). Transformation to a reference frame moving with the speed c (i.e. (x,t) is replaced by $(x-ct,t)$, and subsequent rescaling readily establishes the equivalence of (1.1) and (1.3). Although equation (1.1) now bears the name KdV, it was apparently first obtained by Boussinesq (1877) (see Miles (1970), Pego and Weinstein (1997) and Nekorkin and Velarde (2002) [see page 20, Fig. 2.1 for a reproduction of the Boussinesq reference] for historical discussions on the KdV equation).

Korteweg and de Vries found the solitary wave solutions (1.2) and, importantly, they showed that they are the limiting members of a two-parameter family of periodic travelling wave solutions, described by elliptic functions and commonly called cnoidal waves,

$$A = b + a\,\mathrm{cn}^2(\gamma(x - Vt); m),$$
$$\text{where} \quad V = 6b + 4(2m - 1)\gamma^2, \, a = 2m\gamma^2. \tag{1.4}$$

Here $\mathrm{cn}(x; m)$ is the Jacobian elliptic function of modulus m ($0 < m < 1$). As $m \to 1$, $\mathrm{cn}(x; m) \to \mathrm{sech}(x)$ and then the cnoidal wave (1.4) becomes the solitary wave (1.2), now riding on a background level b. On the other hand, as $m \to 0$, $\mathrm{cn}(x; m) \to \cos 2x$ and so the cnoidal wave (1.4) collapses to a linear sinusoidal wave (note that in this limit $a \to 0$).

This solitary wave solution found by Korteweg and de Vries had earlier been obtained directly from the governing equations (in the absence of surface tension) independently by Boussinesq (1871, 1877)and Rayleigh (1876) who were motivated to explain the now very well-known observations and experiments of Russell (1844). Curiously, it was not until quite recently that it was recognised that the KdV equation is not strictly valid if surface tension is taken into account and $0 < W < 1/3$, as then there is a resonance between the solitary wave and very short capillary waves.

After this ground-breaking work of Korteweg and de Vries, interest in solitary water waves and the KdV equation declined until the dramatic discovery of the *soliton* by Zabusky and Kruskal in 1965. Through numerical integrations of the KdV equation they demonstrated that the solitary wave (1.2) could be generated from quite general initial conditions, and could survive intact collisions with other solitary waves, leading them to coin the term soliton. Their remarkable discovery, followed almost immediately by the theoretical work of Gardner, Greene, Kruskal and Miura (1967) showing that the KdV equation was — *integrable* through an inverse scattering transform, led to many other startling discoveries and marked the birth of soliton theory as we know it today. The implication is that the solitary wave is the key component needed to describe the behaviour of long, weakly nonlinear waves. In particular, a general localized initial condition will lead as $t \to \infty$ the generation of a finite number of solitons and some decaying radiation (see Figure 1). A brief account of soliton theory for the KdV equation is in section 5.

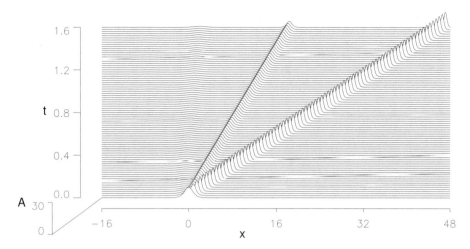

Figure 1. The generation of three solitons from a localized initial condition for the KdV equation (1.1).

1.3   Various extensions to the canonical KdV model

The KdV equation is uni-directional. A two-dimensional version of the KdV equation is the KP equation (Kadomtsev and Petviashvili, 1970),

$$(A_t + 6AA_x + A_{xxx})_x \pm A_{yy} = 0. \tag{1.5}$$

This equation includes the effects of weak diffraction in the y-direction, in that y-derivatives scale as ϵ^2 whereas x-derivatives scale as ϵ. Like the KdV equation it is an integrable equation. When the "+"-sign holds in (1.5), this is the KPII equation, and it can be shown that then the solitary wave (1.2) is stable to transverse disturbances. On the other hand if the "−"-sign holds, this is the KPI equation for which the solitary wave is unstable; instead this equation supports "lump" solitons. Both KPI and KPII are integrable equations. To take account of stronger transverse effects, and/or to allow for bi- directional propagation in the x-direction it is customary to replace the KdV equation with a Boussinesq system of equations; these combine the long wave approximation to the dispersion relation with the leading order nonlinear terms and occur in several asymptotically equivalent forms. Another modification of the KP equation occurs when it is necessary to take account of background rotation, leading to the rotation-modified KP equation (see, for instance, Grimshaw 2001), in which a term $\mp f^2 A$ is added to the left-hand side of equation (1.5), where f is measure of the background rotation.

Although the KdV equation (1.1) is historically associated with water waves, it occurs in many other physical contexts, where it can be derived by an asymptotic multi-scale reduction from the relevant governing equations. Typically the outcome is

$$A_t + cA_x + \mu AA_x + \lambda A_{xxx} = 0. \tag{1.6}$$

Here c is the relevant linear long wave speed for the mode whose amplitude is $A(x,t)$, while μ and λ, the coefficients of the quadratic nonlinear and linear dispersive terms respectively, are determined from the properties of this same linear long wave mode and, like c depend on the particular physical system being considered. Note that the linearization of (1.6) has the linear dispersion relation $\omega = ck - \lambda k^3$ for linear sinusoidal waves of frequency ω and wavenumber k; this expression is just the truncation of the full dispersion relation for the wave mode being considered, and immediately identifies the origin of the coefficient λ. Similarly, the coefficient μ can be identified with the an amplitude-dependent correction to the linear wave speed. Transformation to a reference frame moving with a speed c and subsequent rescaling shows that (1.6) can be transformed to the canonical form (1.1). Specifically, let

$$\mu A = 6U\tilde{A}, \quad x - ct = \left(\frac{\lambda}{U}\right)^{1/2} \tilde{x}, \quad t = \left(\frac{\lambda}{U^3}\right)^{1/2} \tilde{t}. \tag{1.7}$$

Here U is a constant velocity scaling factor inserted to make the transformed variables dimensionless; a convenient choice is often $U = |c|$ provided $c \neq 0$. Then, after removing the superscript, equation (1.6) collapses to the canonical form (1.1). Equations of the form (1.6) arise in the study of internal solitary waves in the atmosphere and ocean, mid-latitude and equatorial planetary waves, plasma waves, ion-acoustic waves, lattice waves, waves in elastic rods and in many other physical contexts (see, for instance, Ablowitz and Segur 1981, Dodd et al 1982, Drazin and Johnson 1989, and Grimshaw 2001). In section 2 we give a brief outline of the derivation of (1.6) for surface and internal waves.

However, an important issue concerning the strict validity of the asymptotic expansion leading to the KdV equation (1.6) is that there should no resonance between the linear long wave mode with speed c and any other part of the linear spectrum of the system being considered. There is an implicit assumption in deriving (1.6) that all solutions, including the solitary wave, are spatially localized. Since we can assume that in the far-field of any solution linearized dynamics apply, this means that we can use the dispersion relation $\omega = \omega(k)$ of the full linearized system for sinusoidal waves of wavenumber k and frequency ω to test whether there is indeed spatial localization. Since all solutions of (1.6) travel with a speed close to the linear long wave speed c, spatial localization requires that there be no resonance between c and any linear phase speed $C(k) = \omega(k)/k$; that is, there are no real solutions for any finite real-valued non-zero k of the resonance condition $c = C(k)$. For water waves without surface tension this condition is satisfied, since the phase speed $C(k)$ is given by

$$C^2(k) = \frac{g \tanh kh}{k},$$

and deceases monotonically from the linear long wave speed $c = C(0) = (gh)^{1/2}$ as k increases from zero to infinity. However, if surface tension is included and the Weber number is such that $0 < W < 1/3$ then a resonance occurs between a long gravity wave and a short capillary wave. In this case, the full system cannot support a spatially localized solitary wave and instead there exist *generalized* solitary waves (see, Boyd 1998 and Grimshaw and Iooss 2003 for general accounts of such waves). They have a central core, described by the KdV solitary wave (refsol) for small amplitudes, but in the far-field have co-propagating non-decaying oscillations with a wavenumber approximately given by the resonance condition. The amplitudes of these oscillations are exponentially small relative to the amplitude of the central core, and this is why any multi-scale asymptotic expansion leading to the KdV equation (1.6) cannot find them. In the case of water waves with surface tension, the resonance arises because the graph of $C(k)$, given now by

$$C^2(k) = \frac{g(1 + Wk^2h^2)\tanh kh}{k},$$

is not monotonic when $0 < W < 1/3$ as it is for zero surface tension $W = 0$, although we note that when $W > 1/3$ the graph is again monotonic and there again exist genuine solitary waves. A similar situation arises for internal waves, where the resonance arises because the dispersion relation is multi-valued (see Aklyas and Grimshaw, 1992). For the first internal mode, there is no resonance and genuine solitary waves exist, but for all higher internal modes there is a resonance between the linear long wave speed c of that mode with the phase speed $C(k)$ of a lower mode, thus leading to generalized solitary waves.

In some physical situations, it is necessary to extend the KdV equation (1.6) with a higher-order cubic nonlinear term of the form $\Sigma A^2 A_x$. After transformation and rescaling, the amended equation (1.6) can be transformed to the so-called Gardner equation

$$A_t + 6AA_x + 6\beta A^2 A_x + A_{xxx} = 0. \tag{1.8}$$

Like the KdV equation, the Gardner equation is integrable by the inverse scattering transform. Here the coefficient β can be either positive or negative, and the structure of the solutions depends crucially on which sign is appropriate. The solitary wave solutions are given by (Kakutani and Yamasaki (1978), Gear and Grimshaw (1983)),

$$A = \frac{a}{b + (1 - b)\cosh^2 \gamma(\theta - V\tau)}, \tag{1.9}$$

$$\text{where} \quad V = a(2 + \beta a) = 4\gamma^2, \quad b = \frac{-\beta a}{(2 + \beta a)}. \tag{1.10}$$

There are two cases to consider. If $\beta < 0$, then there is a single family of solutions such that $0 < b < 1$ and $a > 0$. As b increases from 0 to 1, the amplitude a increases from 0 to a maximum of $-1/\beta$ while the speed V also increases from 0 to a maximum of $-1/\beta$. In the limiting case when $b \to 1$ the solution (1.9) describes the so-called "thick" solitary wave, which has a flat crest of amplitude $a_m = -1/\beta$ (see Figure 2)

For the case when $\beta > 0$, $b < 0$ and there are two families of solitary waves. One is defined by $-1 < b < 0$, has $a > 0$, and as b decreases from 0 to -1, the amplitude a increases from 0 to ∞, while the speed V also increases from 0 to ∞. The other is defined by $-\infty < b < -1$, has $a < 0$ and, as b increases from $-\infty$ to -1, the amplitude a decreases from $-2/\beta$ to ∞. In the limit $b \to -1$

$$A = a\,\text{sech}\,2\gamma(\theta - V\tau), \quad V = \beta a^2 = 4\gamma^2$$

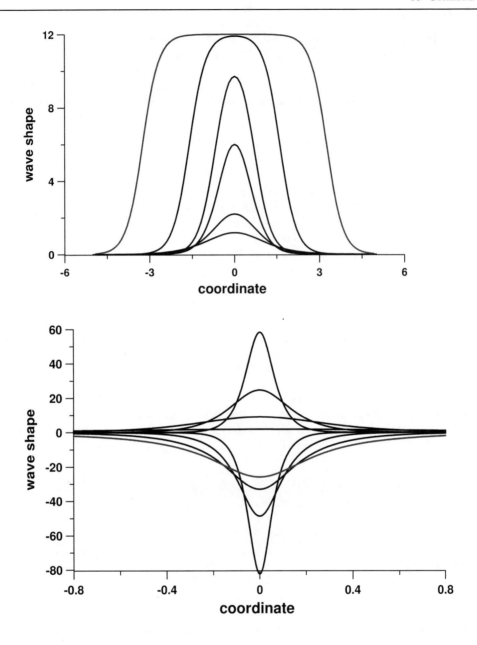

Figure 2. Solitary wave solutions of the extended KdV equation (1.8); upper panel for $\beta < 0$; lower panel for $\beta > 0$.

where here a an take either sign. On the other hand, as $b \to -\infty$, $\gamma \to 0$ and the solitary wave (1.9) reduces to the algebraic form

$$A = \frac{a_0}{1 + \beta a_0^2 \theta^2 / 4}, \quad a_0 = -\frac{2}{\beta}.$$

2 Asymptotic derivation for surface and internal waves

2.1 Formulation from the two-dimensional Euler equations

Here we shall give an outline of the derivation of the KdV equation for surface and internal waves (a more complete discussion can be found in the review articles by Grimshaw 2001, Holloway et al 2001 and Rottman and Grimshaw 2001, on which this present account is based). Thus we consider an inviscid, incompressible fluid which is bounded above by a free surface and below by a flat rigid boundary. We shall suppose that the flow is two-dimensional and can be described by the spatial coordinates (x, z) where x is horizontal and z is vertical (see Figure 3). This configuration is appropriate for the modelling of internal solitary waves in coastal seas, and also in straits, fjords or lakes provided that the effect of lateral boundaries can be ignored. With some modifications this model can also be used to describe atmospheric solitary waves.

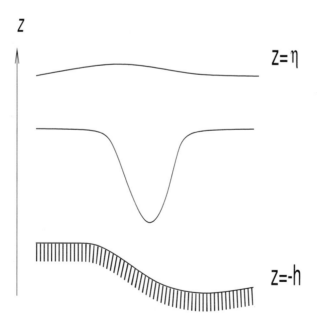

Figure 3. Coordinate system.

Here, in the basic state the fluid has density $\rho_0(z)$, a corresponding pressure $p_0(z)$ such that $p_{0z} = -g\rho_0$ describes the basic hydrostatic equilibrium, and a horizontal shear flow

$u_0(z)$ in the x−direction. Then, in standard notation, the equations of motion relative to this basic state are

$$\rho_0(u_t + u_0 u_x + w u_{0z}) + p_x = -\rho_0(u u_x + w u_z)$$
$$-\rho(u_t + u_0 u_x + w u_{0z} + u u_x + w u_z), \quad (2.1)$$
$$p_z + g\rho = -(\rho_0 + \rho)(w_t + u_0 w_x + u w_x + w w_z), \quad (2.2)$$
$$g(\rho_t + u_0 \rho_x) - \rho_0 N^2 w = -g(u \rho_x + w \rho_z), \quad (2.3)$$
$$u_x + w_z = 0. \quad (2.4)$$

Here $(u_0 + u, w)$ are the velocity components in the (x, z) directions, $\rho_0 + \rho$ is the density, $p_0 + p$ is the pressure and t is time. $N(z)$ is the bouyancy frequency, defined by

$$\rho_0 N^2 = -g\rho_{0z}. \quad (2.5)$$

The boundary conditions are

$$w = 0 \quad \text{at} \quad z = -h, \quad (2.6)$$
$$p_0 + p = 0, \quad \text{at} \quad z = \eta, \quad (2.7)$$
$$w = \eta_t + u_0 \eta_x + u \eta_x \quad \text{at} \quad z = \eta. \quad (2.8)$$

Here, the fluid has undisturbed constant depth h, and η is the displacement of the free surface from its undisturbed position $z = 0$. Note that the effect of the earth's rotation has been neglected. We shall proceed to obtain the KdV equation for internal solitary waves, but note that to recover the theory for water waves from this general formulation it is sufficient just to put the density $\rho_0(z) = \text{constant}$.

Thus to describe internal solitary waves we seek solutions whose horizontal length scales are much greater than h, and whose time scales are much greater than N^{-1}. We shall also assume that the waves have small amplitude. Then the dominant balance is obtained by equating to zero the terms on the left-hand side of (2.1 - 2.4); together with the linearization of the free surface boundary conditions we then obtain the set of equations describing linear long wave theory. To proceed it is useful to use the vertical particle displacement ζ as the primary dependent variable. It is defined by

$$\zeta_t + u_0 \zeta_x + u \zeta_x + w \zeta_z = w. \quad (2.9)$$

Note that it then follows that the perturbation density field is given by $\rho = \rho_0(z - \zeta) - \rho_0(z) \approx \rho_0 N^2 \zeta$ as $\zeta \to 0$, where we have assumed that as $x \to -\infty$, the density field relaxes to its basic state. The isopycnal surfaces (i.e. $\rho_0 + \rho = \text{constant}$) are then given by $z = z_0 + \zeta$ where z_0 is the level as $x \to -\infty$. In terms of ζ, the kinematic boundary condition (2.8) becomes simply $\zeta = \eta$ at $z = \eta$.

Linear long wave theory is now obtained by omitting the right-hand side of equations (2.1 - 2.4), and simultaneously linearizing boundary conditions (2.7,2.8). Solutions are sought in the form

$$\zeta = A(x - ct)\phi(z), \quad (2.10)$$

while the remaining dependent variables are then given by

$$u = (c - u_0)A\phi_z, \quad p = \rho_0(c - u_0)^2 A\phi_z, \quad \rho = \rho_0 N^2 A\phi. \quad (2.11)$$

Here c is the linear long wave speed, and the modal functions $\phi(z)$ are defined by the boundary-value problem,

$$\{\rho_0(c - u_0)^2 \phi_z\}_z + \rho_0 N^2 \phi = 0, \quad \text{in} \quad -h < z < 0, \tag{2.12}$$

$$\phi = 0 \quad \text{at} \quad z = -h, \qquad (c - u_0)^2 \phi_z = g\phi \quad \text{at} \quad z = 0 \tag{2.13}$$

Typically, the boundary-value problem (2.12, 2.13) defines an infinite sequence of modes, $\phi_n^{\pm}(z)$, $n = 0, 1, 2, \ldots$, with corresponding speeds $c_n^{\pm}$. Here, the superscript "$\pm$" indicates waves with $c_n^{+} > u_M = \max u_0(z)$ and $c_n^{-} < u_M = \min u_0(z)$ respectively. We shall confine our attention to these regular modes, and consider only stable shear flows. Nevertheless, we note that there may also exist singular modes with $u_m < c < u_M$ for which an analogous theory can be developed (Maslowe and Redekopp, 1980). Note that it is useful to let $n = 0$ denote the surface gravity waves for which c scales with $\sqrt{gh}$, and then $n = 1, 2, 3, \ldots$ denotes the internal gravity waves for which c scales with Nh. In general, the boundary-value problem (2.12, 2.13) is readily solved numerically. Typically, the surface mode ϕ_0 has no extrema in the interior of the fluid and takes its maximum value at the surface $z = 0$, while the internal modes $\phi_n^{\pm}(z)$, $n = 1, 2, 3, \ldots$, have n extremal points in the interior of the fluid, and vanish near $z = 0$ (and, of course, also at $z = -h$).

It can now be shown that, within the context of linear long wave theory, any localised initial disturbance will evolve into a set of outwardly propagating modes, each given by an expression of the form (2.10). Indeed, it can be shown that the solution of the linearised long wave equations is given asymptotically by

$$\zeta \sim \sum_{n=0}^{\infty} A_n^{\pm}(x - c_n^{\pm} t)\phi_n^{\pm}(z), \quad \text{as} \quad t \to \infty. \tag{2.14}$$

Here the amplitudes $A_n^{\pm}(x)$ are determined in terms of the initial conditions,

$$\zeta = \zeta^{(0)}(x, z), \quad u = u^{(0)}(x, z), \quad \text{at} \quad t = 0, \tag{2.15}$$

by the integral expressions,

$$I_n^{\pm} A_n^{\pm}(x) = \int_{-h}^{0} \rho_0\{(c_n^{\pm} - u_0)\zeta_z^{(0)} + u^{(0)} + u_{0z}\zeta^{(0)}\}\phi_{nz}^{\pm}\, dz, \tag{2.16}$$

$$\text{where} \qquad I_n^{\pm} = 2 \int_{-h}^{0} \rho_0(c_n^{\pm} - u_0)\phi_{nz}^{\pm 2} dz. \tag{2.17}$$

Assuming thats the speeds $c_n^{\pm}$ of each mode are sufficiently distinct, it is sufficient for large times to consider just a single mode. Henceforth, we shall omit the indices and assume that the mode has speed c, amplitude A and modal function $\phi(z)$. Then, as time increases, the hitherto neglected nonlinear terms begin to have an effect, and cause wave steepening. However, this is opposed by the terms representing linear wave dispersion, also neglected in the linear long wave theory. A balance between these two effects emerges as time increases and the outcome is the KdV equation for the wave amplitude.

2.2   Aysmptotic expansion

The formal derivation of the evolution equation requires the introduction of two small parameters, α and ϵ, respectively characterising the wave amplitude and dispersion. The KdV balance requires $\alpha = \epsilon^2$, with a corresponding timescale of ϵ^{-3}. The asymptotic analysis required is well understood (e.g. Benney (1996), Lee and Beardsley (1974), Ostrovsky (1978), Maslowe and Redekopp (1980), Grimshaw (1981a), Tung *et al* (1981)), so we shall give only a brief outline here. We introduce the scaled variables

$$T = \epsilon\alpha t , \quad X = \epsilon(x - ct) , \tag{2.18}$$

and then put

$$\zeta = \alpha A(X,T)\phi(z) + \alpha^2\zeta_2 + \dots , \tag{2.19}$$

with similar expressions analogous to (2.11) for the other dependent variables. At leading order, we get the linear long wave theory for the modal function $\phi(z)$ and the speed c, defined by the modal equations (2.12, 2.13). Since the modal equations are homogeneous, we are free to impose a normalization condition on $\phi(z)$. A commonly used condition is that $\phi(z_m) = 1$ where $|\phi(z)|$ achieves a maximum value at $z = z_m$. In this case the amplitude αA is uniquely defined as the amplitude of ζ (to $0(\alpha)$) at the depth z_m.

Then, at the next order, we obtain the equation for ζ_2,

$$\{\rho_0(c - u_0)^2\zeta_{2Xz}\}_z + \rho_0 N^2\zeta_{2X} = M_2, \quad \text{for} \quad -h < z < 0 \quad , \tag{2.20}$$

$$\zeta_{2X} = 0 \quad \text{at} \quad z = -h, \quad \rho_0(c - u_0)^2\zeta_{2Xz} - \rho_0 g\zeta_{2X} = N_2, \quad \text{at} \quad z = 0 \quad . \tag{2.21}$$

Here the inhomogeneous terms M_2, N_2 are known in terms of $A(X,T)$ and $\phi(z)$, and are given by

$$M_2 = 2\{\rho_0(c - u_0)\phi_z\}_z A_T + 3\{\rho_0(c - u_0)^2\phi_z^2\}_z AA_X$$
$$- \rho_0(c - u_0)^2\phi A_{XXX} , \tag{2.22}$$

$$N_2 = 2\{\rho_0(c - u_0)\phi_z\}A_T + 3\{\rho_0(c - u_0)^2\phi_z^2\}AA_X . \tag{2.23}$$

Note that the left-hand side of equations (2.20, 2.21)) are identical to the equations defining the modal function (i.e. (2.12, 2.13)), and hence can be solved only if a certain compatibility condition is satisfied. In general the compatibility condition is that the inhomogenous terms in (2.20, 2.21) should be orthogonal to the solutions of the adjoint of the modal equations (2.12, 2.13). Here, rather than following this general procedure, we obtain the compatibility condition by a direct construction of ζ_2. Thus a formal solution of (2.20) which satisfies the first boundary condition in (2.21) is

$$\zeta_{2X} = A_{2X}\phi + \phi \int_{-h}^{z} \frac{M_2\psi}{W}dz - \psi \int_{-h}^{z} \frac{M_2\phi}{W}dz , \tag{2.24}$$

$$\text{where} \quad W = \rho_0(c - u_0)^2\{\phi_z\psi - \psi_z\phi\} . \tag{2.25}$$

Here $\psi(z)$ is a solution of the modal equation (2.12) which is linearly independent of $\phi(z)$, and so, in particular $\psi(-h) \neq 0$. W (2.25) is the Wronskian of these two solutions, and is a constant independent of z. Indeed, the expression (2.25) can be used to obtain

ψ explicitly in terms of ϕ. The homogeneous part $A_{2X}\phi$ of the expression (2.24) for ζ_{2X} introduces the second-order amplitude $A_2(X, T)$. Next, we insist that the expression (2.24) for ζ_{2X} should satisfy the second boundary condition in (2.21). The result is the compatibility condition

$$\int_{-h}^{0} M_2\phi \, dz = [N_2\phi]_{z=0} \,. \tag{2.26}$$

Note that the amplitude A_2 is left undetermined at this stage.

Substituting the expressions (2.22, 2.23) into (2.26) we obtain the required evolution equation for A, namely the KdV equation

$$A_T + \mu A A_X + \lambda A_{XXX} = 0 \,. \tag{2.27}$$

Taking into account the scaling (2.18) this is just (1.6), where here the coefficients μ and λ are given by

$$I\mu = 3\int_{-h}^{0} \rho_0(c - u_0)^2 \phi_z^3 \, dz \,, \tag{2.28}$$

$$I\lambda = \int_{-h}^{0} \rho_0(c - u_0)^2 \phi^2 \, dz \,, \tag{2.29}$$

$$\text{where} \quad I = 2\int_{-h}^{0} \rho_0(c - u_0)\phi_z^2 \, dz \,. \tag{2.30}$$

Here I is just $I_n^{\pm}$ (2.17) with the subscript and superscript omitted. The KdV equation (2.27) is to be solved with the initial condition $A(X, T = 0) = A_0(X)$ where $A_0(X)$ is determined from the linear long wave theory, that is, it is given by (2.16), and is in essence the projection of the original initial conditions (2.15) onto the relevant mode. As mentioned in section 1 and described in more detail in section 5, localized initial conditions lead to the generation of a finite number of solitary waves.

Confining attention to waves propagating to the right, so that $c > u_M = \max u_0(z)$, we see that I and λ are always positive. For the surface mode, $\phi_z > 0$ and $\phi(0) = 1$ so we see that $\mu > 0$. Further, recalling that for the internal modes the modal functions are normalised so that $\phi(z_m) = 1$ where z_m is an extremal point, then it is readily shown that for the usual situation of a near-surface pycnocline, μ is negative for the first internal mode. However, in general μ can take either sign, and in some special situations may even be zero. Explicit evaluation of the coefficients μ and λ requires knowledge of the modal function, and hence they are usually evaluated numerically.

To illustrate the procedure, consider first the case of *water waves*. We put the density $\rho = $ constant so that then $N^2 = 0$ (2.5). Then

$$\phi = \frac{z + h}{h} \quad \text{for} \quad -h < z < 0, \quad c = (gh)^{1/2} \,. \tag{2.31}$$

$$\text{and so} \quad \mu = \frac{3c}{2h}, \quad \lambda = \frac{ch^2}{6}, \tag{2.32}$$

in agreement with (1.3) when the Weber number $W = 0$.

Similarly, for *interfacial waves*, let the density be a constant ρ_1 in an upper layer of height h_1 and $\rho_2 > \rho_1$ in the lower layer of height $h_2 = h - h_1$. That is

$$\rho_0(z) = \rho_1 H(z + h_1) + \rho_2 H(-z - h_1),$$

$$\text{so that} \quad \rho_0 N^2 = g(\rho_2 - \rho_1)\delta(z + h_1).$$

Here $H(z)$ is the Heaviside function and $\delta(z)$ is the Dirac δ-function. For simplicity, we shall also replace the free boundary with a rigid boundary so that the upper boundary condition for $\phi(z)$ becomes just $\phi(0) = 0$. This is a good approximation for oceanic internal solitary waves.

Then we find that

$$\phi = \frac{z + h}{h_2} \quad \text{for} - h < z < h_1,$$

$$\phi = -\frac{z}{h_1} \quad \text{for} - h_1 < z < 0,$$

$$c^2 = \frac{g(\rho_2 - \rho_1)h_1 h_2}{\rho_1 h_2 + \rho_1 h_2} \tag{2.33}$$

Substitution into (2.28) and (2.29) yields

$$\mu = \frac{3c(\rho_2 h_1^2 - \rho_1 h_2^2)}{h_1 h_2(\rho_2 h_1 + \rho_1 h_2)},$$

$$\lambda = \frac{c h_1 h_2(\rho_2 h_2^2 + \rho_1 h_1^2)}{(\rho_2 h_1 + \rho_1 h_2)} \tag{2.34}$$

Note that for the usual oceanic situation when $\rho_2 - \rho_1 << \rho_2$, the nonlinear coefficient μ for these interfacial waves is negative when $h_1 < h_2$ (that is, the interface is closer to the free surface than the bottom), and is positive in the reverse case. The case when $h_1 \approx h_2$ leads to the necessity to use the extended KdV equation (2.35).

2.3   Higher-order models and other extensions

Proceeding to the next highest order will yield an equation set analogous to (2.20, 2.21) for ζ_3, whose compatibility condition then determines an evolution equation for the second-order amplitude A_2. We shall not give details here, but note that using the transformation $A + \alpha A_2 \to A$, and then combining the KdV equation (2.27) with the evolution equation for A_2 will lead to a higher-order KdV equation for A,

$$A_T + \mu A A_X + \lambda A_{XXX}$$

$$+\alpha\{\lambda_1 A_{XXXXX} + \Sigma A^2 A_X + \mu_1 A A_{XXX} + \mu_2 A_X A_{XX}\} = 0.$$

Explicit expresions for the coefficients are given by Gear and Grimshaw (1983), Lamb and Yan (1996), and Grimshaw *et al* (2002)). However, this equation is not unique, as the near-identity transformation $A \to A + \alpha(aA^2 + bA_{XX})$ asymptotically reproduces the same equation but with altered coefficients,

$$(\lambda_1, \Sigma, \mu_1, \mu_2) \to (\lambda_1, \Sigma - a\mu, \mu_1, \mu_2 - 6a\lambda + 2b\mu).$$

Note that to be Hamiltonian, $\mu_2 = 2\mu_1$. Further the enhanced transformation

$$A \to A + \alpha(aA^2 + bA_{XX} + a'A_X \int^X A\,dX + b'XA_T),$$

can asymptotically reduce the higher-order equation to the KdV equation provided that $\mu \neq 0, \lambda \neq 0$.

A particularly impotant special case of the higher-order KdV equation arises when the nonlinear coefficient μ (2.27) in the KdV equation is close to zero. In this situation, the cubic nonlinear term in the higher-order KdV equation is the most important higher-order term. The KdV equation (2.27) may then be replaced by the extended KdV equation,

$$A_T + \mu A A_X + \alpha \Sigma A^2 A_X + \lambda A_{XXX} = 0. \tag{2.35}$$

For $\mu \approx 0$, a rescaling is needed and the optimal choice is to assume that μ is $0(\epsilon)$, and then replace A with A/ϵ. In effect the amplitude parameter is ϵ in place of ϵ^2. The coefficient Σ of the cubic nonlinear term is given in terms of integrals of the modal function ϕ and the second order correction χ_2 (see Grimshaw et al 2002 for details).

In some atmospheric and oceanic applications, the depth h is not necessarily small relative to the horizontal length scale of the solitary wave, but nevertheless the density stratification is effectively confined to a thin layer of depth h_1, which is much shorter that the horizontal length scales. In this case, a different theory is needed, and was first developed by Benjamin (1967) and Davis and Acrivos (1967). Several variants are possible, so, to be specific, we shall describe an oceanic case when $\rho_0(z)$, and $u_0(z)$ vary only in a near-surface layer of depth h_1, below which $\rho_0(z) = \rho_\infty$ (a constant) and $u_0(z) = 0$, while the ocean bottom is now given by $z = -H/\epsilon$ (i.e. $H = \epsilon h$). The modal function is again defined by (2.12, 2.13) but the boundary condition at the bottom is now replaced by a matching condition that $\phi_z \to 0$ as $z \to -\infty$. To derive the evolution equation, we again use the asymptotic expansion (2.12) but now with $\alpha = \epsilon$ and restricted to the near-surface layer. This expansion is matched to an appropriate solution in the deep-fluid region where Laplace's equation holds at leading order. The outcome is the intermediate long-wave (ILW) equation (Kubota $et\ al$ (1978), Maslowe and Redekopp (1980), Grimshaw (1981a), Tung $at\ al$ (1981),

$$A_T + \mu A A_X + \delta\ \mathbf{L}(A_X)) = 0, \tag{2.36}$$

$$\text{where} \qquad \mathbf{L}(A) - \frac{1}{2\pi} \int_{-\infty}^{\infty} k \coth kH \exp(ikX)\ \mathbf{F}(A)dk,$$

$$\text{and} \qquad \mathbf{F}(A) \int_{-\infty}^{\infty} A \exp(-ikX))dX.$$

Here the nonlinear coefficient μ is again given by (2.28) with $-h$ now replaced by $-\infty$, while the dispersive coefficient δ is defined by $I\delta = (\rho_0 c^2 \phi^2)_{z\to-\infty}$. In the limit $H \to \infty$, $k \coth kH \to |k|$ on the integrand and (2.36) becomes the Benjamin-Ono (BO) equation. In the opposite limit $H \to 0$, (2.36) reduces to the KdV equation.

An important variant of the ILW equation (2.36) arises when it is supposed that the deep ocean is infinitely deep ($H \to \infty$) and weakly stratified, with a constant buoyancy frequency ϵN_0. Then the operator $\mathbf{L}(A)$ in (2.36) is replaced by (see Maslowe and Redekopp 1980, Grimshaw 1981b)

$$\mathbf{L}_m(A) = -\frac{1}{2\pi} \int_{-\infty}^{\infty} (k^2 - m^2)^{\frac{1}{2}} \exp(ikX)\, \mathbf{F}(A)dk\,, \tag{2.37}$$

where $m = N_0/c$. Now internal gravity waves can propagate vertically in the deep fluid region, and to ensure that these waves are outgoing, a radiation condition is needed. Thus $(k^2 - m^2)^{\frac{1}{2}}$ is either real and positive for $k^2 > m^2$, or $i\,sign\,k(m^2 - k^2)^{\frac{1}{2}}$ for $k^2 < m^2$. As $m \to 0$, (2.37) becomes the BO equation.

3 Solitary waves in a variable environment

3.1 Variable-coefficient KdV equation

In many physical situations it is necessary to take account of the fact that solitary waves propagate through a variable environment. This means that the coefficients c, μ and λ in (1.6) are functions of x, while an additional term $c(Q_x/2Q)A$ needs to be included, where $Q(x)$ is a magnification factor. Thus (1.6) is replaced by

$$A_t + cA_x + c\frac{Q_x}{2Q}A + \mu AA_x + \lambda A_{xxx} = 0\,. \tag{3.1}$$

After transforming to new variables,

$$X = (\int^x dx/c) - t\,, \qquad B = Q^{1/2}A\,, \tag{3.2}$$

the variable-coefficient KdV equation is obtained for $B(x, X)$,

$$B_x + \nu(x)BB_X + \delta(x)B_{XXX} = 0\,. \tag{3.3}$$

$$\text{where} \qquad \nu = \mu/cQ^{1/2}, \delta = \lambda/c^3.$$

It is assumed here that

$$\frac{\partial}{\partial x} << \frac{\partial}{\partial X}\,.$$

In general, equation (3.3) is not an integrable equation and must be solved numerically, although we shall exhibit some asymptotic solutions below.

3.2 Fission of a solitary wave

There are two distinct limiting situations in which some analytical progress can be made. First, let it be supposed that the coefficients $\nu(x), \delta(x)$ in (3.3) vary rapidly with respect to the wavelength of a solitary wave, and consider then the case when these coefficients make a rapid transition from the values ν_-, δ_- in $x < 0$ to the values ν_+, δ_+ in $x > 0$. Then a steady solitary wave can propagate in the region $x < 0$, given by

$$B = b\,\mathrm{sech}^2(\gamma(X - Wx)) \quad \text{where} \quad W = \frac{\nu_- b}{3} = 4\delta_-\gamma^2\,. \tag{3.4}$$

It will pass through the transition zone $x \approx 0$ essentially without change. However, on arrival into the region $x > 0$ it is no longer a permissible solution of (3.3), which now has constant coefficients ν_+, δ_+. Instead, with $x = 0$, the expression (3.4) now forms an effective initial condition for the new constant-coefficient KdV equation. Using the spectral problem (5.1) and the inverse scattering transform, the solution in $x > 0$ can now be constructed; indeed in this case the spectral problem (5.1) has an explicit solution (e.g. Drazin and Johnson, 1989). The outcome is that the initial solitary wave *fissions* into N solitons, and some radiation. The number N of solitons produced is determined by the ratio of coefficients $R = \nu_+ \delta_- / \nu_- \delta_+$. If $R > 0$ (i.e. there is no change in polarity for solitary waves), then $N = 1 + [((8R+1)^{1/2} - 1)/2]$ ([$\cdots$] denotes the integral part); as R increases from 0, a new soliton (initially of zero amplitude) is produced as R successively passes through the values $m(m + 1)/2)$ for $m = 1, 2, \cdots$. But if $R < 0$ (i.e. there is a change in polarity) no solitons are produced and the solitary wave decays into radiation.

For instance, for water waves, $c = (gh)^{1/2}, \mu = 3c/2h, \lambda = ch^2/6, Q = c$ and so $\nu = 3/(2hc^{1/2}), \delta = h^2/(6c^2)$ where h is the water depth. It can then be shown that a solitary water wave propagating from a depth h_- to a depth h_+ will fission into N solitons where N is given as above with $R = (h_-/h_+)^{9/4}$; if $h_- > h_+, N \geq 2$, but if $h_- > h_+$ then $N = 1$ and no further solitons are produced (Johnson 1973).

3.3 Slowly-varying solitary wave

Next, consider the opposite situation when the coefficients $\nu(x), \delta(x)$ in (3.3) vary slowly with respect to with respect to the wavelength of a solitary wave. In this case a multi-scale perturbation technique (see Grimshaw 1979, or Grimshaw and Mitsudera 1993) can be used in which the leading term is

$$B \sim b \operatorname{sech}^2 \gamma (X - \int_{x_0}^{x} W \, dx), \tag{3.5}$$

$$\text{where} \quad W = \frac{\nu b}{3} = 4\delta \gamma^2. \tag{3.6}$$

Here the wave amplitude $b(x)$, and hence also $W(x), \gamma(x)$, are slowly-varying functions of x. Their variation is most readily determined by noting that the variable-coefficient KdV equation (3.3) possesses a conservation law,

$$\int_{-\infty}^{\infty} B^2 dX = \text{constant}. \tag{3.7}$$

which expresses conservation of wave-action flux. Substitution of (3.5) into (3.7) gives

$$\frac{2b^2}{3\gamma} = \text{constant}, \quad \text{so that} \quad b = \text{constant}(\frac{\delta}{\nu})^{1/3}. \tag{3.8}$$

This is an explicit equation for the variation of the amplitude $b(x)$ in terms of $\nu(x), \delta(x)$.

However, the variable-coefficient KdV equation (3.3) also has a conservation law for mass,

$$\int_{-\infty}^{\infty} B dX = \text{constant}. \tag{3.9}$$

Thus, although the slowly-varying solitary wave conserves wave-action flux it cannot simultaneously conserve mass. Instead, it is accompanied by a trailing shelf of small amplitude but long length scale given by B_s, so that the conservation of mass gives

$$\int_{-\infty}^{\phi} B_s \, dX + \frac{2b}{\gamma} = \text{constant}\,,$$

where $\phi = \int_{x_0}^{x} W \, dx$ ($X = \phi$ gives the location of the solitary wave) and the second term is the mass of the solitary wave (3.5). Differentiation then yields the amplitude $B_- = v_s(X = \phi)$ of the shelf at the rear of the solitary wave,

$$B_- = \frac{3\gamma_x}{\nu\gamma^2}\,. \tag{3.10}$$

This shows that if the wavenumber γ increases (decreases) as the solitary wave deforms, then the trailing shelf amplitude B_- has the same (opposite) polarity to the solitary wave. Once B_- is known the full shelf $B_s(X, x)$ is found by solving (3.3) with the boundary condition that $B_s(X = \phi) = B_-$ (see El and Grimshaw (2002), where it is shown that the trailing shelf may eventually generate secondary solitary waves).

A simple application of this asymptotic theory shows that, for a solitary water wave propagating over a variable depth $h(x)$, the amplitude varies as h^{-1}. The trailing shelf has positive (negative) polarity relative to the wave itself according as $h_x < (>)0$.

A situation of particular interest occurs if the coefficient $\nu(x)$ changes sign at some particular location (note that in most physical systems the coefficient δ of the linear dispersive term in (3.3) does not vanish for any x). This commonly arises for internal solitary waves in the coastal ocean, where typically in the deeper water, $\nu < 0, \delta > 0$ so that internal solitary waves propagating shorewards are waves of depression. But in shallower water, $\nu > 0$ and so only internal solitary waves of elevation can be supported. The issue then arises as to whether an internal solitary wave of depression can be converted into one or more solitary waves of elevation as the critical point, where ν changes sign, is traversed. This problem has been intensively studied (see, for instance, Grimshaw et al (1998) and the references therein), and the solution depends on how rapidly the coefficient ν changes sign. If ν passes through zero rapidly compared to the local width of the solitary wave, then the solitary wave is destroyed, and converted into a radiating wavetrain (see the discussion above in the first paragraph of this section). On the other hand, if ν changes sufficient slowly for the present theory to hold (i.e. (3.8) applies), we find that as $\nu \to 0$ then $B \to 0$ in proportion to $|\nu|^{1/3}$, while $B_- \to \infty$ as $|\nu|^{-8/3}$. Thus, as the solitary wave amplitude decreases, the amplitude of the trailing shelf, which has the opposite polarity, grows indefinitely until a point is reached just prior to the critical point where the slowly-varying solitary wave asymptotic theory fails. A combination of this trailing shelf and the distortion of the solitary wave itself then provide the appropriate "initial" condition for one or more solitary waves of the opposite polarity to emerge as the critical point is traversed. However, it is clear that in situations, as here, where $\nu \approx 0$, it will be necessary to include a cubic nonlinear term in (3.3), thus converting it into a variable-coefficient Gardner equation (cf. (1.8)), that is,

$$B_x + \nu(x)BB_X + \chi(x)B^2 B_X + \delta(x)B_{XXX} = 0\,. \tag{3.11}$$

This case has been studied by Grimshaw, Pelinovsky and Talipova (1999), who showed that the outcome depends on the sign of the coefficient (χ) of the cubic nonlinear term in (3.11) at the critical point. If $\chi > 0$ so that solitary waves of either polarity can exist when $\nu = 0$, then the solitary wave preserves its polarity (i.e. remains a wave of depression) as the critical point is traversed. On the other hand if $\chi < 0$ so that no solitary wave can exist when $\chi = 0$ then the solitary wave of depression may be converted into one or more solitary waves of elevation.

4 The generation of solitary waves by flow over topography

4.1 Undular bores

To this point, we have considered only the situation when solitary waves arise as a solutions of the initial value problem for the KdV equation (1.1). This is appropriate in many cases. For instance, in the ocean or atmosphere, or in laboratory experiments, internal solitary waves can be generated when the pycnocline is given a localized displacement. This, as described in section 2, provides an initial condition for the KdV. When this initial displacement has the correct polarity, it will evolve into a finite number of solitary waves. This scenario often occurs in the coastal oceans when the barotropic ocean tide interacts with the continental shelf to generate an internal tide, which provides the necessary pycnocline displacement. This, in turn, evolves into propagating internal solitary waves.

However, KdV solitary waves can also arise by direct forcing mechanisms. A common such situation is when a fluid flow interacts with a localized topographic obstacle in a situation of near criticality. Here a critical flow is one which supports a linear long wave of zero speed. In this case, the waves generated are unable to escape the vicinity of the obstacle, and hence can be said to be directly forced. The inability of the waves to propagate away rapidly from the forcing region means that nonlinearity is needed from the outset. As we will show below the KdV equation (1.1) is then replaced by a forced KdV equation, whose principal solutions resemble "undular bores". Hence, we first provide a description of the "undular bore".

The term "undular bore" is widely used in the literature in a variety of contexts and several different meanings. Here, we need to make it clear that we are concerned with non-dissipative flows, in which case an undular bore is intrinsically unsteady. In general, an undular bore is an oscillatory transition between two different basic states. A simple representation of an undular bore can be obtained from the solution of the KdV equation (1.1) with the initial condition that

$$A = A_0 H(-x), \tag{4.1}$$

where we assume at first that $A_0 > 0$. Here $H(x)$ is the Heaviside function (i.e. $H(x) = 1$ if $x > 0$ and $H(x) = 0$ if $x < 0$). The solution can in principle be obtained through the inverse scattering transform. However, it is more instructive to use the asymptotic method developed by Gurevich and Pitaevskii (1974), and Whitham (1974). In this approach, the solution of (1.1) with this initial condition is represented as the modulated periodic wave train (1.4) supplemented here with a mean term d, that is

$$A = a\{b(m) + \text{cn}^2(\gamma(x - Vt); m) + d, \tag{4.2}$$

$$\text{where} \quad b = \frac{1 - m}{m} - \frac{E(m)}{mK(m)}, \quad a = 2m\gamma^2,$$

$$\text{and} \quad V = 6d + 2a\left\{\frac{2 - m}{m} - \frac{3E(m)}{mK(m)}\right\}. \tag{4.3}$$

We recall from section 1 that as the modulus $m \to 1$, this becomes a solitary wave, but as $m \to 0$ it reduces to sinusoidal waves of small amplitude.

The asymptotic method of Gurevich and Pitaevskii (1974) and Whitham (1974) is to let the expression (4.2) describe a modulated periodic wavetrain in which the amplitude a, the mean level d, the speed V and the wavenumber γ are all slowly varying functions of x and t. The relevant asymptotic solution corresponding to the initial condition (4.1) can now be constructed in terms of the similarity variable x/t, and is given by

$$\frac{x}{t} = 2A_0\left\{1 + m - \frac{2m(1 - m)(K(m)}{E(m) - (1 - m)K(m)}\right\},$$

$$\text{for} \quad -6A_0 < \frac{x}{t} < 4A_0, \tag{4.4}$$

$$a = 2A_0 m, \quad d = A_0\left\{m - 1 + \frac{2E(m)}{K(m)}\right\}. \tag{4.5}$$

Ahead of the wavetrain where $x/t > 4A_0, A = 0$ and at this end, $m \to 1$, $a \to 2A_0$ and $d \to 0$; the leading wave is a solitary wave of amplitude $2A_0$ relative to a mean level of 0. Behind the wavetrain where $x/t < -6A_0$, $A = A_0$ and at this end $m \to 0$, $a \to 0$, and $d \to A_0$; the wavetrain is now sinusoidal with a wavenumber γ given by $6\gamma^2 \approx A_0$. Further, it can be shown that on any individual crest in the wavetrain, $m \to 1$ as $t \to \infty$. In this sense, the undular bore evolves into a train of solitary waves.

If $A_0 < 0$ in the initial condition (4.2), then an "undular bore" solution analogous to that described by (4.2, 4.4) does not exist. Instead, the asymptotic solution is a rarefraction wave,

$$A = 0 \quad \text{for} \quad x > 0,$$

$$A = \frac{x}{6t} \quad \text{for} \quad A_0 < \frac{x}{6t} < 0,$$

$$A = A_0, \quad \text{for} \quad \frac{x}{6t} < A_0(< 0). \tag{4.6}$$

Small oscillatory wavetrains are needed to smooth out the discontinuities in A_x at $x = 0$ and $x = -6A_0$ (for further details, see Gurevich and Pitaevskii 1974).

4.2   Forced KdV equation

The generation of an undular bore requires an initial condition in which $A \to A_\pm$ with $A_- > A_+$ as $x \to \pm\infty$; note that (4.1) is the simplest such condition. A common situation where this type of initial condition may be generated occurs when a steady transcritical flow encounters a localized topographic obstacle, in the context of the flow of a density-stratified fluid as described in section 2. Here a flow $u_0(z)$ is said to be critical if it can support a wave mode whose speed $c \approx 0$, in the frame of reference of the topographic obstacle. Let us suppose that the bottom boundary of the stratified fluid is given by $z = -h + \alpha^2 F(X)$, where $X = \epsilon x$, $F(X)$ is spatially localized, and for a KdV balance $\alpha = \epsilon^2$, as before. Let the speed $c = \alpha\Delta$ where Δ is a detuning parameter. Then it was shown by Grimshaw and Smyth (1986) that the KdV equation (2.27) is replaced by the forced KdV (fKdV) equation

$$A_T + \Delta A_X + \mu A A_X + \lambda A_{XXX} + \Gamma F_X(X) = 0 \,, \tag{4.7}$$

$$\text{where} \qquad I\Gamma = -2\rho_0 u_0 \phi_z(z = -h) \,.$$

Here the coefficients μ, λ, I are given by (2.28, 2.29, 2.30) respectively with $c = 0$. Without loss of generality we shall suppose that the oncoming flow is left to right so that $u_0 > 0$; indeed, it is sufficient to assume that I (2.30)< 0. In this case $\lambda < 0, I < 0$, and $\Delta > 0(< 0)$ defines supercritical (subcritical) flow respectively. Also it then follows that $\mu < 0(> 0)$ for a solitary wave of elevation (depression). The fKdV equation (4.7) has been derived in several other physical contexts, and is a canonical model equation to describe transcritical flow interaction with an obstacle. It was first derived in the context of water waves (for instance, for transcritical free surface flows generated by a moving pressure forcing by Akylas (1985), or by flow over an obstacle by Cole (1985)), and indeed equation (4.7) can describe that case as well by choosing the barotropic mode $n = 0$, or more simply by setting the density $\rho_0 = $ constant.

As for the KdV equation (2.27) we may now rescale the fKdV equation (4.7) into a canonical form. Let

$$-\mu\epsilon^2 A \;=\; 6U\tilde{A} \,, \quad X = \epsilon \left(\frac{-\lambda}{U}\right)^{1/2} \tilde{x} \,, \quad T = \epsilon^3 \left(\frac{-\lambda}{U^3}\right)^{1/2} \tilde{t} \,,$$

$$\epsilon^2 \Delta \;=\; U\tilde{\Delta} \,, \quad \epsilon^4 \mu \Gamma F = 6U^2 \tilde{F} \,. \tag{4.8}$$

Here, as before, U is a constant scaling factor inserted to make the transformed variables dimensionless; here a convenient choice would be $U = u_0(0)$. After removing the superscript, equation (4.7) collapses to the canonical fKdV equation

$$-A_t - \Delta A_x + 6AA_x + A_{xxx} + F_x(x) = 0 \,. \tag{4.9}$$

This is to be solved with the initial condition that $A(x,0) = 0$, which corresponds to a slow introduction of the topographic obstacle. An important issue here is the polarity of the forcing in (4.9), that is, whether it has positive (negative) polarity $F(x) \geq 0(\leq 0)$. Referring to the scaling (4.8) we see that positive polarity in the original dimensional coordinates leads to positive (negative) polarity in the dimensionless equation (4.9) according as $\mu\Gamma > 0(< 0)$. Note that for a uniform flow $u_0(z) = U > 0$, for a surface wave both $\mu > 0$ and $\Gamma > 0$ while for an interfacial wave μ may be either positive or negative depending on whether the interface is closer to the bottom, and again $\Gamma > 0$ (see 2.34).

A typical solution of (4.9) is shown in Figure 4 for exact criticality, when $\Delta = 0$ and the obstacle provides a positive, and isolated, forcing term. That is $F(x)$ is positive,

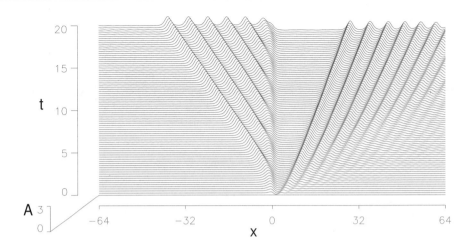

Figure 4. The solution of the fKdV equation (4.9) at exact criticality, $\Delta = 0$.

and non-zero only in a vicinity of $x = 0$, with a maximum value of $F_M > 0$. The solution is characterised by upstream and downstream wavetrains connected by a locally steady solution over the obstacle. For supercritical flow ($\Delta < 0$) the upstream wavetrain weakens, and for sufficiently large $|\Delta|$ detaches from the obstacle, while the downstream wavetrain intensifies and for sufficiently large $|\Delta|$ forms a stationary lee wave field. On the other hand, for supercritical flow ($\Delta > 0$) he upstream wavetrain develops into well-separated solitary waves while the downstream wavetrain weakens and moves further downstream (for more details see Grimshaw and Smyth 1986 and Smyth 1987). The origin of the upstream and downstream wavetrains can be found in the structure of the locally steady solution over the obstacle. In the transcritical regime this is characterised by a transition from a constant state A_- upstream of the obstacle to a constant state A_+ downstream of the obstacle, where $A_- < 0$ and $A_+ > 0$. It is readily shown that $\Delta = 3(A_+ + A_-)$ independently of the details of the forcing term $F(x)$. Explicit determination of A_+ and A_- requires some knowledge of the forcing term $F(x)$. However, in the "hydraulic" limit when the linear dispersive term in (3.20a) can be neglected, it is readily shown that

$$6A_{\pm} = \Delta \mp (12F_M)^{1/2}. \tag{4.10}$$

This expression also serves to define the transcritical regime, which is

$$|\Delta| < (12F_M)^{1/2}. \tag{4.11}$$

Thus upstream of the obstacle there is a transition from the zero state to A_-, while downstream the transition is from A_+ to 0; each transition is effectively generated at $X = 0$.

Both transitions are resolved by "undular bore" solutions as described above. That in $x < 0$ is exactly described by (4.2) to (4.5) with x replaced by $\Delta t - x$, and A_0 by A It occupies the zone

$$\Delta - 4A_- < \frac{x}{t} < \max\{0, \ \Delta + 6A_-\}. \tag{4.12}$$

Note that this upstream wavetrain is constrained to lie in $x < 0$, and hence is only fully realised if $\Delta < -6A_-$. Combining this criterion with (4.10) and (4.11)) defines the regime

$$-(F_M)^{1/2} < \Delta < -\frac{1}{2}(F_M)^{1/2}, \tag{4.13}$$

where a fully developed undular bore solution can develop upstream. On the other hand, the regime $\Delta < -6A_-$ or

$$-\frac{1}{2}(F_M)^{1/2} < \Delta < (F_M)^{1/2}, \tag{4.14}$$

is where the upstream undular bore is only partially formed, and is attached to the obstacle. In this case the modulus m of the Jacobian elliptic function varies from 1 at the leading edge (thus describing solitary waves) to a value m_- (< 1) at the obstacle, where m_- can be found from (4.4) by replacing x with Δ and A_0 with A_-.

The transition in $x > 0$ can also be described by (4.2) to (4.5) where we now replace x with $(\Delta + 6A_+)t - x$, A_0 with $-A_+$, and d with $d - A_+$. This "undular bore" solution occupies the zone

$$\max\{0, \ \Delta - 2A_+\} < \frac{x}{t} < \Delta - 12A_+. \tag{4.15}$$

Here, this downstream wavetrain is constrained to lie in $x > 0$, and hence is only fully realised if $\Delta > 2A_+$. Combining this criterion with (4.10) and (4.11) defines the regime (4.14), and so a fully detached downstream undular bore coincides with the case when the upstream undular bore is attached to the obstacle. On the other hand, in the regime (4.13), when the upstream undular bore is detached from the obstacle, the downstream undular bore is attached to the obstacle, with a modulus m_+(< 1) at the obstacle, where m_+ can be founding from (4.4) by replacing x with $\Delta - 6A_+$ and A_0 with A_+. Indeed now a stationary lee wavetrain develops just behind the obstacle (for further details, see Smyth, 1987).

For the case when the obstacle has negative polarity (that is $F(x)$ is negative, and non-zero only in the vicinity of $x = 0$), the upstream and downstream solutions are qualitatively similar to those described above for positive forcing. However, the solution in the vicinity of the obstacle remains transient, and this causes a modulation of the "undular bore" solutions.

5 Soliton Theory

The remarkable discovery of Gardner, Greene, Kruskal and Miura (1967) that the KdV equation was integrable through an inverse scattering transform marked the beginning of soliton theory. Their pioneering work was followed by the work of Zakharov and Shabat (1972) which showed that another well-known nonlinear wave equation, the nonlinear

Schrödinger equation, was also integrable by an inverse scattering transform. This result that the integrability of the KdV equation was not an isolated result, was followed closely by analogous results for the modified KdV equation (Wadati, 1972) and the Sine-Gordon equation (Ablowitz et al, 1973). In 1974, Ablowitz, Kaup, Newell and Segur provided a generalisation and unification of these results in the AKNS scheme. From this point there has been an explosive and rapid development of soliton theory in many directions (see, for instance, Ablowitz and Segur 1981, Dodd et al 1982, Newell 1985 and Drazin and Johnson 1989)

For the KdV equation (1.1) the starting point is the Lax pair (Lax 1968) for an auxiliary function $\phi(x, t)$,

$$L\phi \equiv -\phi_{xx} - u\phi = \lambda\phi, \tag{5.1}$$

$$\phi_t = B\phi \equiv (u_x + C)\phi + (4\lambda - 2u)\phi_x. \tag{5.2}$$

Here $C(t)$ depends on the normalization of ϕ. The first of these equations (5.1), with suitable boundary conditions at infinity (see below) defines a spectral problem for ϕ in the spatial variable x with a spectral parameter λ, and with the time variable t as a parameter. The second equation (5.2) then describes how the spectral function ϕ evolves in time. If it is now assumed that λ is independent of time (i.e. $\lambda_t = 0$ then the KdV equation (1.1) is just the compatibility condition for these two equations (5.1, 5.2); that is, it emerges as a result of the condition that $(\phi_{xx})_t = (\phi_t)_{xx}$. In terms of the operators L, B defined in the Lax pair (5.1,5.2) the KdV equation can be written in the symbolic form $L_t = BL - LB$ (Lax, 1968). This form indicates the path to further generalizations, in that other nonlinear wave equations can be obtained by choosing different operators L, B.

The general strategy for integration of the KdV equation now consists of three steps. Here we will describe the process under the hypothesis that we seek solutions $u(x, t)$ of the KdV equation (1.1) which decay to zero sufficiently fast as $x \to \pm\infty$ and have the initial condition, $u(x, 0) = u_0(x)$. First, we insert the initial condition into the spectral problem (5.1) to obtain obtain the scattering data (these will be defined precisely below). Then (5.2) is used to move the scattering data forward in time; it transpires that is a very simple process, and note in particular that the spectral parameter λ is independent of t and hence is determined by the initial condition. The third step is to invert the scattering data at time $t > 0$ and so recover $u(x, t)$; this is the most difficult step, but for the KdV equation can be reduced to solution of a linear integral equation. Thus, the three steps constitute a linear algorithm for the solution of the KdV equation, and it is in this sense that it is said that the Lax pair (5.1,5.2) constitute integrability of the KdV equation.

The spectral problem (5.1) for the KdV equation consists of two parts. The *discrete* spectrum is found by seeking solutions such that $\phi \to 0$ as $x \to \pm\infty$, which requires that $\lambda < 0$. It can be shown that there then exist a *finite* set of discrete eigenvalues $\lambda = -\kappa_n^2, n = 1, 2, \cdots, N$, and corresponding real eigenfunctions ϕ_n such that

$$\phi_n \sim c_n \exp\left(-\kappa_n x\right) \qquad \text{as} \quad x \to \infty. \tag{5.3}$$

There is a similar condition as $x \to -\infty$, namely that $\phi_n \sim d_n \exp(\kappa_n x)$. The real constants c_n, d_n are determined once the normalization condition is satisfied, that is,

$$\int_{-\infty}^{\infty} \phi_n^2 \, dx = 1 \,. \tag{5.4}$$

The continuous spectrum consists of all $\lambda > 0$, and so we set $\lambda = k^2$ where k is real. Then we define the scattering problem for solutions $\phi(x; k)$ of (5.1) by the boundary conditions,

$$\phi \sim \exp(-ikx) + R(k) \exp(ikx) \qquad \text{as} \quad x \to \infty \,, \tag{5.5}$$

$$\phi \sim T(k) \exp(-ikx) \qquad \text{as} \quad x \to -\infty \,. \tag{5.6}$$

The scattering data then consists of the set $(\kappa_n, c_n, n = 1, 2, \cdots, N)$ together with the reflection coefficient $R(k)$. It is useful to note that $R(k)$ may be continued into the upper half of the complex k-plane, has there a set of simple poles at $k = i\kappa_n$, and $R \to 1$ as $|k| \to \infty$.

The next step is to determine from (5.2) how the scattering data evolves in time (note that the dependence on time t has been suppressed in the preceding paragraph). First, we recall that the discrete eigenvalues κ_n are independent of t. Next, we multiply (5.2) by ϕ_n and integrate the result over all x; on also using (5.1) it is readily found that

$$\frac{d}{dt} \int_{-\infty}^{\infty} \phi_n^2 \, dx = C_n \int_{-\infty}^{\infty} \phi_n^2 \, dx \,,$$

where here the constant C in (5.2) must be indexed with n to become C_n. But then the normalization condition (5.4) implies that $C_n = 0$. Now substitute (5.3) into (5.2 to show that

$$\frac{dc_n}{dt} = 4\kappa_n^3 c_n \qquad \text{so that} \quad c_n(t) = c_n(0) \exp(\kappa_n^3 t) \,. \tag{5.7}$$

For the continuous spectrum, the asymptotic expressions (5.5, 5.6) are substituted into (5.2). Now it is found that the constant $C(k) = 4ik^3$, and that

$$\frac{dR}{dt} = 8ik^3 R \qquad \text{so that} \quad R(k; t) = R(0; k) \exp(8ik^3 t) \,, \tag{5.8}$$

Similarly, it can be shown that $T(k; t) = T(k; 0)$.

The final step is the inversion of the scattering data at time t to recover the potential $u(x, t)$ in (5.1). This is accomplished through the Marchenko integral equation for the function $K(x, y)$

$$K(x, y) + F(x + y) + \int_x^{\infty} K(x, z) \, F(y + z) \, dz = 0 \,. \tag{5.9}$$

Here the function $F(x)$ is known in terms of the scattering data at time t,

$$F(x) = \sum_{n=1}^{N} c_n^2(t) \exp(-\kappa_n x) + \frac{1}{2\pi} \int_{-\infty}^{\infty} R(k; t) \exp(ikx) \, dk \,. \tag{5.10}$$

Here the t-dependence of K, F has been suppressed as the linear integral equation (5.9) is solved with t fixed. Then

$$u(x,t) = 2\frac{\partial}{\partial x}\{K(x,x;t)\}, \qquad (5.11)$$

where the t-dependence in K has been restored.

The inverse scattering transform described by (5.9,5.10) enables one to find the solution of the KdV equation (1.1) for an arbitrary localised initial condition. The most important outcome is that as $t \to \infty$, the solution evolves into N rank-ordered solitons propagating to the right $(x > 0)$, and some decaying radiation propagating to the left $(x < 0)$,

$$u \sim \sum_{n=1}^{N} 2\kappa_n^2 \text{sech}^2(\kappa_n(x - 4\kappa_n^2 t - x_n)) + \text{radiation}. \qquad (5.12)$$

Here the N solitons derive directly from the discrete spectrum, where each eigenvalue $-\kappa_n$ generates a soliton of amplitude $2\kappa_n^2$, while the phase shifts x_n are determined from the constants $c_n(0)$. The continuous spectrum is responsible for the decaying radiation, which decays at each fixed $x < 0$ as $t^{-1/3}$.

The important special case when the reflection coefficient $R(k) \equiv 0$ leads to the N-soliton solution, for which there bis no radiation. Indeed, the N- soliton solution can be obtained as an explicit solution of the Marchenko equation (5.9). We illustrate the procedure for $N = 1, 2$. First, for $N = 1$, $F(x) = c^2 \exp \kappa(x - 4\kappa^2 t)$ where we have omitted the subscript $n = 1$ for simplicity. Then seek a solution of (5.9) in the form $K(x,y,t) = L(x,t) \exp(-\kappa y)$, where L can be found by simple algebra. The outcome is that

$$L(x,t) = \frac{-2\kappa c(0)^2 \exp(-\kappa x + 8\kappa^2 t)}{2\kappa + c(0)^2 \exp(-2\kappa x + 8\kappa^2 t)}.$$

Finally u is found from (5.11),

$$u = 2\kappa^2 \text{sech}^2(\kappa(x - 4\kappa^2 t) - x_1).$$

This is just the solitary wave (1.2) of amplitude $2\kappa^2$; the phase shift x_1 is such that $c(0)^2 = 2\kappa \exp(2\kappa x_1)$. The procedure for $N = 2$ follows a similar course. Thus, with $R \equiv 0, N = 2$ in F (5.10), seek a solution of the Marchenko equation (5.9) in the form $K(x,y,t) = L_1(x,t) \exp(-\kappa_1 y) + L_2(x,t) \exp(-\kappa_2 y)$, and again $L_{1,2}$ can be found by simple algebra. The outcome is the 2-soliton solution. For instance, with $\kappa_1 = 1, \kappa_2 = 2$, this is

$$u = 12\frac{3 + 4\cosh(2x - 8t) + \cosh(4x - 64t)}{[3\cosh(x - 28t) + \cosh(3x - 36t)]^2}. \qquad (5.13)$$

It can be readily shown that

$$u \sim 8\text{sech}^2(2(x - 16t \mp x_2) + 2\text{sech}^2(x - 4t \mp x_1) \qquad \text{as} \quad t \to \pm\infty, \qquad (5.14)$$

where the phase shifts $x_{1,2} = (-1/2, 1/4) \ln 3$. Thus the 2-soliton solution describes the elastic collision of two solitons, in which each survives the interaction intact, and the only memory of the collision are the phase shifts; note that $x_1 < 0, x_2 > 0$, so that the larger soliton is displaced forward and the smaller soliton is displaced backward. The general case of an N-soliton is analogous and is essentially a sequence of pair-wise 2-soliton interactions.

The integrability of the KdV equation (1.1) is also characterized by the existence of an infinite set of independent conservation laws. The most transparent conservation laws are

$$\int_{-\infty}^{\infty} u \, dx = \text{constant}, \tag{5.15}$$

$$\int_{-\infty}^{\infty} u^2 \, dx = \text{constant}, \tag{5.16}$$

$$\int_{-\infty}^{\infty} u^3 - \frac{1}{2} u_x^2 \, dx = \text{constant}, \tag{5.17}$$

which may be associated with the conservation of mass, momentum and energy resepctively. Indeed (5.15) is obtained from the KdV equation (1.1) by integrating over x, while (5.16,5.17) are obtained in an analogous manner after first multiplying (1.1) by u, u^2 respectively. However, it transpires that these are just the first three conservation laws in an infinite set, where each successive conservation law contains a higher power of u than the preceding one. This may be demonstrated using the inverse scattering transform (see Ablowitz and Segur 1981, Dodd et al 1982, and Newell 1985). However, here we use the original method based on the Miura transformation as adapted by Gardner. The Miura transformation is

$$u = -v_x - v^2, \tag{5.18}$$

$$v_t + 6v^2 v_x + v_{xxx} = 0. \tag{5.19}$$

Here (5.19) is the modified KdV equation. Direct substitution of (5.18) into the KdV equation shows that if v solves the modified KdV equation (5.19), then u solves the KdV equation (1.1). This discovery was the starting point for the discovery of the inverse scattering transform, since if one considers (5.18) as an equation for v and writes $v = \phi_x/\phi$, followed by a Galilean transformation for u (i.e. $u \to u - \lambda, x \to x - 6\lambda t$), one obtains the spectral problem (5.1). Here we follow a different route, and write

$$v = \frac{1}{2\epsilon} - \epsilon w,$$

which (after a shift $x \to x + 3t/2\epsilon^2$) converts the mKdV equation (5.19) into the Gardner equation (1.8)with $\beta = -\epsilon^2$. Apart from a constant, which may be removed by a Galilean transformation, the corresponding expression for u is the Gardner transformation,

$$u = w + \epsilon w_x - \epsilon^2 w^2. \tag{5.20}$$

Thus if w solves the Gardner equation

$$w_t + 6ww_x - 6\epsilon^2 w^2 w_x + w_{xxx} = 0, \tag{5.21}$$

then u solves the KdV equation (1.1).

Next, we observe that the Gardner equation (5.21) has the conservation law

$$\int_{-\infty}^{\infty} w \, dx = \text{constant}. \tag{5.22}$$

Since $w \to u$ as $\epsilon \to 0$, we write the formal asymptotic expansion

$$w \sim \sum_{n=0}^{\infty} \epsilon^n w_n.$$

It follows from (5.22) that then

$$\int_{-\infty}^{\infty} w_n \, dx = \text{constant},$$

for each $n = 0, 1, 2, \cdots$. But substitution of this same asymptotic expansion for w into (5.20) generates a sequence of expressions for w_n in terms of u, of which the first few are

$$w_0 = u, \qquad w_1 = -u_x, \qquad w_2 = -u^2 + u_{xx}.$$

Thus we see that $n = 0, 2$ give the conservation laws (5.15,5.16) respectively, while $n = 1$ is an exact differential. It may now be shown that all even values of n yield non-trivial and independent conservation laws, while all odd values of n are exact differentials.

The KdV equation belongs to a class of nonlinear wave equations, which have Lax pairs and are integrable through an inverse scattering transform. It shares with these equations several other remarkable featurea, such as the Hirota bilinear form, Bäcklund transforms and the Painlevé property. Detailed descriptions of these and other properties of the KdV equation can be found in the referenced texts.

Bibliography

M.J. Ablowitz, D.J. Kaup,, A.C. Newell, and H. Segur Method for solving the Sine-Gordon equation. *Phys. Lett.* 30: 1262-1264, 1973.

M.J. Ablowitz,, D.J. Kaup,, A.C. Newell, and H. Segur. The inverse scattering transform - Fourier analysis for nonlinear problems. *Studies in Applied Mathematics* 53: 249-315, 1974

M.J. Ablowitz, and H. Segur. *Solitons and the inverse scattering transform*. SIAM Studies in Applied Mathematics 4, SIAM, Philadelphia, 1981.

T.R. Akylas and R. Grimshaw. Solitary internal waves with oscillatory tails. *J.Fluid Mech.*, 242: 279-298, 1992.

T.B. Benjamin, 1967. Internal waves of permanent form in fluids of great depth. *J. Fluid Mech.*, 29: 559-592, 1967.

D.J. Benney. Long non-linear waves in fluid flows. *J. Math. Phys.*, 45: 52-63, 1966.

J.P. Boyd. Weakly Nonlinear Solitary Waves and Beyond-All-Orders Asymptotics. *Kluwer*. Boston, 1998.

M.J. Boussinesq. Theórie de l'intumescence liquide appellée onde solitaire ou de translation, se propageant dans un canal rectangulaire. *Comptes Rendus Acad. Sci (Paris)* 72: 755-759, 1871.

M.J. Boussinesq. Essai sur la theórie des eaux courantes, *Mémoires présentées par diverse savants à l'Académie des Sciences Inst. France (series 2)* 23: 1-680, 1877.

R.E. Davis and A. Acrivos. Solitary internal waves in deep water. *J. Fluid Mech.*, 29: 593-607, 1967.

R.K. Dodd, J.C. Eilbeck, J.D. Gibbon and H.C. Morris. *Solitons and nonlinear wave equations.* Academic, London. 1982.

P.G. Drazin and R.S. Johnson. *Solitons: an Introduction.* Cambridge University Press, Cambridge, 1989.

G.A. El and R. Grimshaw. Generation of undular bores in the shelves of slowly-varying solitary waves. *Chaos*, 12: 1015-1026, 2002.

C.S. Gardner, J.M. Greene, M.D. Kruskal and R.M. Miura. Method for solving the Korteweg-de Vries equation. *Physical Review Letters*, 19: 1095-1097, 1967.

J. Gear and R. Grimshaw. A second-order theory for solitary waves in shallow fluids. *Phys. Fluids,* 26: 14-29, 1993.

R. Grimshaw. Slowly varying solitary waves. Korteweg-de Vries equation. *Proc. Roy. Soc.*, 368A: 359-375, 1979.

R. Grimshaw. Evolution equations for long nonlinear waves in stratified shear flows. *Stud. Appl. Maths.*, 65: 159-188, 1981a.

R. Grimshaw, 1981b. Slowly varying solitary waves in deep fluids. *Proc. Roy. Soc.*, 376A: 319-332, 1981b.

R. Grimshaw, 2001. Internal solitary waves. In *Environmental Stratified Flows*, Kluwer, Boston, Chapter 1: 1-28, 2001.

R. Grimshaw and G. Iooss. Solitary waves of a coupled Korteweg-de Vries system. *Math. and Computers in Simulation*, 62: 31–40, 2003.

R. Grimshaw, and H. Mitsudera. Slowly-varying solitary wave solutions of the perturbed Korteweg-de Vries equation revisited. *Stud. Appl. Math.*, 90: 75-86, 1993.

R. Grimshaw, E. Pelinovsky and T. Talipova 1998. Solitary wave transformation due to a change in polarity. *Stud. Appl. Math.*, 101: 357-388.

R. Grimshaw, E. Pelinovsky and T. Talipova. Solitary wave transformation in a medium with sign-variable quadratic nonlinearity and cubic nonlinearity. *Physica D*, 132: 40-62, 1999.

R. Grimshaw,, E. Pelinovsky and O. Poloukhina. Higher-order Korteweg-de Vries models for internal solitary waves in a stratified shear flow with a free surface. *Nonlinear Processes in Geophysics*, 9: 221-235, 2002.

A.V. Gurevich, A.V. and L.P. Pitaevskii. Nonstationary structure of a collisionless shock wave. *Sov. Phys. JETP*, 38: 291-29, 1974.

P. Holloway, E. Pelinovsky and T. Talipova 2001. Internal tide transformation and oceanic internal solitary waves. In *Environmental Stratified Flows* Kluwer, Boston, Chapter 2: 29-60, 2001

R.S. Johnson. On the development of a solitary wave moving over an uneven bottom. *Proc. Camb. Phil. Soc.*, 73: 183-203, 1973.

B.B. Kadomtsev, and V.I. Petviashvili. On the stability of solitary waves in weakly dispersive media. *Soviet Physics Doklady*, 15: 539-541, 1970.

T. Kakutani, and N. Yamasaki. Solitary waves on a two-layer fluid. *J. Phys. Soc. Japan.* 45: 674-679, 1978.

D.J. Korteweg, and H. de Vries. On the change of form of long waves advancing in a rectangular canal, and on a new type of long stationary waves. *Philosophical Magazine*, 39: 422-443, 1895.

T. Kubota, D.R.S. Ko and L.S. Dobbs. Weakly nonlinear long internal gravity waves in stratified fluids of finite depth. *AIAA J. Hydronautics*, 12: 157-168, 1978.

K.G. Lamb and L. Yan. The evolution of internal wave undular bores: comparisons of a fully nonlinear numerical model with weakly-nonlinear theory. *J.Phys. Ocean.*, 99: 843-864, 1996.

P.D. Lax. Integrals of nonlinear equations of evolution and solitary waves. *Comm. Pure Appl. Math*, 21: 467-490, 1968.

C.Y. Lee, and R.C. Beardsley. The generation of long nonlinear internal waves in a weakly stratified shear flow. *J.Geophys, Res.*, 79: 453-462, 1974.

S.A. Maslowe and L.G. Redekopp. Long nonlinear waves in stratified shear flows. *J. Fluid Mech.*, 101: 321-348, 1980.

J.W. Miles. Solitary waves *Annual Review of Fluid Mechanics* 12: 11- 43, 1980.

V. I. Nekorkin and M. G. Velarde. *Synergetic Phenomena in Active Lattices. Patterns, Waves, Solitons, Chaos*. Springer-Verlag, Berlin, 2002.

A.C. Newell. *Solitons in mathematics and physics*. CBMS-NSF Series in Applied Mathematics 48, SIAM, Philadelphia, 1985.

L.A. Ostrovsky. Nonlinear internal waves in rotating fluids. *oceanology*, 18: 181-191, 1978.

R.L. Pego and M.J. Weinstein. Convective linear stability of solitary waves for Boussinesq equations. *Studies Appl. Math.*, 99: 311-375, 1997.

Lord Rayleigh. On waves. *Phil. Mag.* 1: 257-279, 1876.

J. Rottman and R. Grimshaw, 2001. Atmospheric internal solitary waves. In *Environmental Stratified Flows*, Kluwer, Boston, Chapter 3: 61-88, 2001.

J.S. Russell. *Report on Waves*, 14th meeting of the British Association for the Advancement of Science: 311-390, 1844.

K.K. Tung, D.R.S. Ko and J.J. Chang. Weakly nonlinear internal waves in shear. *Stud. Appl. Math.*, 65: 189-221, 1981.

M. Wadati. The exact solution of the modified Korteweg-de Vries equation. *J. Phys. Soc. Japan*, 32: 62-69, 1972.

G.B. Whitham. *Linear and Nonlinear Waves*. Wiley, New York, 1974.

N.J. Zabusky and M.D. Kruskal. Interactions of solitons in a collisionless plasma and the recurrence of initial states. *Physical Review Letters*, 15: 240-243, 1965.

Weakly Nonlinear Wave Packets and the Nonlinear Schrödinger Equation

Frédéric Dias[*] and Thomas Bridges[†]

[*] Centre de Mathématiques et de Leurs Applications, Ecole Normale Supérieure de Cachan, Cachan, France

[†] Department of Mathematics and Statistics, University of Surrey, Guildford, UK

Abstract. This chapter describes weakly nonlinear wave packets. The primary model equation is the nonlinear Schrödinger (NLS) equation. Its derivation is presented for two systems : the Korteweg–de Vries equation and the water-wave problem. Analytical as well as numerical results on the NLS equation are reviewed. Several applications are considered, including the study of wave stability. The bifurcation of waves when the phase and the group velocities are nearly equal as well as the effects of forcing on the NLS equation are discussed. Finally, recent results on the effects of dissipation on the NLS equation are also given.

1 Introduction

The nonlinear Schrödinger (NLS) equation provides a canonical description for the envelope dynamics of a quasi-monochromatic plane wave (the carrier) propagating in a weakly nonlinear dispersive medium when dissipative processes are negligible. The NLS equation assumes weak nonlinearities but a finite dispersion at the scale of the carrier. Cumulative nonlinear interactions result in a modulation of the wave amplitude on large spatial and temporal scales.

The NLS equation arises in various physical contexts in the description of nonlinear waves such as water waves at the free surface of an ideal fluid or plasma waves. It provides a canonical description of the envelope dynamics of a dispersive wave train

$$\eta = \epsilon A \exp[\mathrm{i}(\mathbf{k} \cdot \mathbf{x} - \omega t)] + \text{c.c.}, \qquad (1.1)$$

with a small ($\epsilon \ll 1$) but finite amplitude, slowly modulated in space and time, propagating in a conservative system. A good physical description is given by Whitham (1974).

Let us consider a nonlinear wave equation written symbolically as

$$\mathcal{L}(\partial_t, \nabla)u + \mathcal{N}(u)u = 0, \qquad (1.2)$$

where $\mathcal{L}$ is a linear operator with constant coefficients and $\mathcal{N}$ a nonlinear function of u and of its derivatives. For a solution of infinitely small amplitude, the nonlinear effects can be neglected, and the equation admits monochromatic wave solutions of the form

$$u = \epsilon A \exp[\mathrm{i}(\mathbf{k} \cdot \mathbf{x} - \omega t)] + \text{c.c.}, \qquad (1.3)$$

with a constant amplitude ϵA. Here c.c. stands for the complex conjugate. The frequency ω and the wave vector $\mathbf{k}$ are linked through the dispersion relation

$$\mathcal{L}(-i\omega, i\mathbf{k}) = 0\,. \tag{1.4}$$

Consider a branch of solution $\omega = \omega(\mathbf{k})$ of (1.4). If we consider a regular perturbation expansion of the solution (1.3), nonlinear effects will accumulate over long times and large propagation distances. Mathematically speaking, resonant terms will be generated in the hierarchy of equations arising at the successive orders, leading to secular terms. One way to deal with these secular terms is the method of multiple scales. The amplitude of the carrier is allowed to vary over slow time and space variables, and its evolution is given by solvability conditions that eliminate the resonances and eventually lead to the NLS equation.

A simple heuristic argument can be given to explain the canonical character of the NLS equation (see for example the monograph by Sulem and Sulem (1999)).

Note that the name "NLS equation" originates from a formal analogy with the Schrödinger equation of quantum mechanics. In this context a nonlinear potential arises in the mean field description of interacting particles.

2 Derivation of the NLS equation

The first derivation of the NLS equation for weakly nonlinear wave packets was apparently given by Benney and Newell (1967). It was re-derived independently by Zakharov (1968). Comprehensive reviews of NLS equations in the context of water waves are given by Peregrine (1983) and by Hammack and Henderson (1993). Craig, Sulem and Sulem (1992) gave a rigorous estimate of the validity of NLS for water waves. But they left open the question of actual convergence of solutions of the water-wave problem to solutions in the form of the modulational ansatz. Some progress was recently made by Schneider (1998), who considered the rigorous derivation of the NLS equation from the Korteweg-de Vries (KdV) equation. A formal derivation of the NLS equation from the KdV equation based on the introduction of multiple scales is presented first. Then two derivations of the NLS equation from the full water-wave equations are given. The first one is based on multiple scale analysis, while the second one is based on Zakharov's integral equation representation of the water-wave problem.

2.1   The method of multiple scales

In order to illustrate the method of multiple scales, we consider the KdV equation, which is the model equation considered in Chapter 1 :

$$u_t + u_{xxx} + 6uu_x = 0\,, \quad (x \in \mathbb{R},\, t \geq 0,\, u(x,t) \in \mathbb{R})\,. \tag{2.1}$$

Its linearization admits monochromatic wave solutions of the form

$$u = \psi \exp[i(kx - \omega t)] + \text{c.c.}\,, \quad \text{with} \ \ \omega = -k^3\,. \tag{2.2}$$

The common ansatz is that the solution has a uniformly valid asymptotic expansion in terms of a small parameter ϵ (the magnitude of the wave amplitude) and that fast and slow variables are introduced. A regular perturbation expansion would lead to the onset of resonant terms resulting from the cumulative effects of weak nonlinearities on long time or large distances.

We look for a solution in the form

$$u = \epsilon(u_0(x, t, X, T, \cdots) + \epsilon u_1(x, t, X, T, \cdots) + \epsilon^2 u_2 + \cdots),$$

where the slow scales $X = \epsilon x, T = \epsilon t$ have been introduced in addition to the fast scales x and t. The new scaled variables are considered as independent. Therefore the linear operator arising in the KdV equation (2.1) can be rewritten as

$$\mathcal{L} = \mathcal{L}^{(0)} + \epsilon \mathcal{L}^{(1)} + \epsilon^2 \mathcal{L}^{(2)} + \cdots, \tag{2.3}$$

where the first three contributions are given by

$$\mathcal{L}^{(0)} = \frac{\partial}{\partial t} + \frac{\partial^3}{\partial x^3}, \tag{2.4}$$

$$\mathcal{L}^{(1)} = \frac{\partial}{\partial T} + 3\frac{\partial^3}{\partial x^2 \partial X}, \tag{2.5}$$

$$\mathcal{L}^{(2)} = \frac{\partial}{\partial \tau} + 3\frac{\partial^3}{\partial x \partial X^2}. \tag{2.6}$$

Note the presence of the additional time scale $\tau = \epsilon^2 t$. At order (ϵ), we have

$$\mathcal{L}^{(0)} u_0 = 0,$$

and we recover the solution (2.2)

$$u_0(x, t) = \psi_0(X, T) \exp[i(kx - \omega t)] + \text{c.c.}$$

of the linear problem (monochromatic wave propagating along the x-axis). The frequency ω is given by the dispersion relation

$$\omega = -k^3. \tag{2.7}$$

At this point, we define the *phase velocity*

$$c = \frac{\omega}{k}.$$

Here $c = -k^2$. It depends on k and there is dispersion. We also define the *phase* $\theta = kx - \omega t$ and the *group velocity*

$$c_g = \frac{d\omega}{dk}.$$

Here $c_g = -3k^2$.

The next step consists in writing the differential equation at order (ϵ^2) :

$$\mathcal{L}^{(0)} u_1 = -\mathcal{L}^{(1)} u_0 - 3\partial_x(u_0^2). \tag{2.8}$$

The following solvability condition is required at order (ϵ^2) :

$$(\partial_T - 3k^2\partial_X)u_0 = 0 \quad \text{or} \quad (\partial_T + c_g\partial_X)u_0 = 0. \tag{2.9}$$

We used the dispersion relation (2.7) to identify $-3k^2$ with the group velocity c_g of the wave packet. On the time scale T, the wave packet is just transported at the group velocity and thus depends only on the variable $\xi = X - c_gT$. We then solve for u_1 :

$$\mathcal{L}^{(0)}u_1 = -3\partial_x(\psi_0^2 e^{2i(kx-\omega t)} + \text{c.c.}). \tag{2.10}$$

The general solution is

$$u_1 = A_2(X,T)e^{2i(kx-\omega t)} + \text{c.c.} + \psi_1 e^{i(kx-\omega t)} + \text{c.c.} + A_0(X,T). \tag{2.11}$$

The term $\psi_1 e^{i(kx-\omega t)}$ can be incorporated into the term $\psi_0 e^{i(kx-\omega t)}$ by defining $\psi = \psi_0 + \epsilon\psi_1$. Then one has

$$A_2 = \frac{-3(2ik)\psi^2}{-2i\omega + (2ik)^3} = \frac{\psi^2}{k^2}.$$

The final step consists in writing the differential equation at order (ϵ^3) :

$$\mathcal{L}^{(0)}u_2 = -\mathcal{L}^{(2)}u_0 - \mathcal{L}^{(1)}u_1 - 6\partial_x(u_0u_1) - 3\partial_X(u_0^2). \tag{2.12}$$

The following solvability condition that eliminates the terms proportional to $e^{i(kx-\omega t)}$ and its complex conjugate is required at order (ϵ^3) :

$$-\partial_\tau\psi - 3ik\partial_{XX}\psi - 6ik|\psi|^2\psi/k^2 - 6ikA_0\psi = 0. \tag{2.13}$$

Equating the coefficients in front of the oscillation free terms gives

$$-\partial_T A_0 - 6\partial_X|\psi|^2 = 0.$$

Taking into account the solvability condition (2.9) yields

$$A_0 = -\frac{2|\psi|^2}{k^2}.$$

Finally, noting that $-6k = \mathrm{d}^2\omega/\mathrm{d}k^2$, (2.13) leads to the cubic NLS equation

$$i\frac{\partial\psi}{\partial\tau} + \frac{1}{2}\frac{\mathrm{d}^2\omega}{\mathrm{d}k^2}\frac{\partial^2\psi}{\partial\xi^2} + \frac{6}{k}|\psi|^2\psi = 0. \tag{2.14}$$

Equation (2.14) describes the time evolution of the wave amplitude from an initial modulation $\psi(\xi, \tau = 0)$.

2.2 Rigorous derivation

In the previous subsection, we provided a formal derivation of the NLS equation. Let

$$\tilde{u} = \epsilon\psi(X,T)\exp[i(kx - \omega t)] + \text{c.c.} + \epsilon^2 A_2(X,T)\exp[2i(kx - \omega t)] + \text{c.c.} + \epsilon^2 A_0(X,T).$$

Then Schneider (1998) proved the following theorem :

Theorem 2.1. *Let $\sigma \geq 3$ and let $\psi \in C([0,T_0], H^{\sigma+5}(\mathbb{R},\mathbb{C}))$ be a solution of the NLS equation (2.14). Then we have $C, \epsilon_0 > 0$ such that for all $\epsilon \in (0, \epsilon_0)$ there are solutions u of the KdV equation (2.1) such that*

$$\sup_{t \in [0, T_0/\epsilon^2]} \|u - \tilde{u}(\psi)\|_{H^\sigma(\mathbb{R},\mathbb{R})} \leq C\epsilon^{3/2}.$$

The proof is not easy. There are two main difficulties : we are dealing with a quasi-linear problem in Sobolev spaces, and the eigenvalues of the linearized problem do not satisfy the non-resonance condition which is needed if quadratic terms are present in the nonlinearity.

2.3 The water-wave problem

In order for this subsection to be self-contained, it is necessary to describe first the water-wave problem (its governing equations, its boundary conditions, its linearization).

The governing equations The three-dimensional flow of an ideal and incompressible fluid is governed by the conservation of mass

$$\nabla \cdot \mathbf{u} = 0 \tag{2.15}$$

and by the conservation of momentum

$$\rho\frac{D\mathbf{u}}{Dt} = \rho\mathbf{g} - \nabla p, \tag{2.16}$$

where the material derivative is defined as $D(*)/Dt \equiv \partial(*)/\partial t + (\mathbf{u} \cdot \nabla)(*)$. The horizontal coordinates are denoted by x and y, and the vertical coordinate by z. The vector $\mathbf{u}(x,y,z,t) = (u,v,w)$ is the velocity field, $\rho(x,y,z)$ is the fluid density, $\mathbf{g}$ is the acceleration due to gravity and $p(x,y,z,t)$ the pressure.

It is assumed below that the density ρ is constant throughout the fluid domain. This assumption is made for the sake of simplicity. A similar analysis can be made for continuously stratified flows. Another assumption which is commonly made to analyze surface waves is that the flow is irrotational (see however the paper by Colin, Dias and Ghidaglia (1995) for rotational effects). If the flow is irrotational ($\nabla \times \mathbf{u} = \mathbf{0}$), there exists a scalar function $\phi(x,y,z,t)$ (the so-called velocity potential) such that $\mathbf{u} = \nabla\phi$. The continuity equation (2.15) becomes

$$\Delta\phi = 0. \tag{2.17}$$

With all these assumptions, the equation of momentum conservation (2.16) can be integrated into the so-called Bernoulli's equation

$$\frac{\partial \phi}{\partial t} + \frac{1}{2}|\nabla \phi|^2 + gz + \frac{p - p_0}{\rho} = 0,\tag{2.18}$$

which is valid everywhere in the fluid. The constant p_0 is a pressure of reference, for example the atmospheric pressure.

The flow domain The surface wave problem consists in solving the incompressible Euler equations (2.15) and (2.16) in a domain bounded above by a free surface (the interface between air and water) and below by a solid boundary (the bottom). The free surface is represented by $F(x, y, z, t) \equiv \eta(x, y, t) - z = 0$. The bottom can be at any depth and have any shape. The driving force is due to gravity, but the effects of surface tension may be equally important in some physical situations.

Boundary conditions Whenever the potential and its derivatives are evaluated on the free surface, the following notation is used :

$$\varphi(x, y, t) = \phi(x, y, \eta, t), \quad \varphi_{(*)}(x, y, t) = \phi_*(x, y, \eta, t),$$

where the star stands for x, y, z or t. Consequently, φ_* and $\varphi_{(*)}$ have different meanings. They are however related since

$$\varphi_x = \varphi_{(x)} + \varphi_{(z)}\eta_x, \quad \varphi_y = \varphi_{(y)} + \varphi_{(z)}\eta_y, \quad \varphi_t = \varphi_{(t)} + \varphi_{(z)}\eta_t.$$

The free surface must be found as part of the solution. Two boundary conditions are required. The first one is the kinematic condition. It can be stated as $DF/Dt = 0$, which leads to

$$\eta_t + \varphi_{(x)}\eta_x + \varphi_{(y)}\eta_y - \varphi_{(z)} = 0.\tag{2.19}$$

In the case of a steady motion, the kinematic condition states that the free surface is a streamline. The second boundary condition is the dynamic condition which states that the forces must be equal on both sides of the free surface. The force normal to the free surface is the difference in pressure and is balanced by the effect of surface tension. If σ denotes the surface tension coefficient and C the curvature of the free surface, $p - p_0 = -\sigma C$. For the air/water interface, $\sigma = 0.074$ N/m. The expression for C is

$$C = \left(\frac{\eta_x}{(1 + \eta_x^2 + \eta_y^2)^{1/2}}\right)_x + \left(\frac{\eta_y}{(1 + \eta_x^2 + \eta_y^2)^{1/2}}\right)_y.\tag{2.20}$$

Bernoulli's equation (2.18) written on the free surface $z = \eta$ gives

$$\varphi_{(t)} + \frac{1}{2}\left(\varphi_{(x)}^2 + \varphi_{(y)}^2 + \varphi_{(z)}^2\right) + g\eta - \frac{\sigma}{\rho}C = 0.\tag{2.21}$$

Finally, the bottom $z = -h(x, y)$ is assumed to be flat : $z = -h$. The boundary condition at the bottom simply is

$$\phi_z(x, y, -h, t) = 0.\tag{2.22}$$

To summarize, one wants to solve for $\eta(x, y, t)$ and $\phi(x, y, z, t)$ the following set of equations :

$$\Delta\phi \equiv \phi_{xx} + \phi_{yy} + \phi_{zz} = 0 \quad \text{for } (x, y, z) \in \mathbb{R} \times \mathbb{R} \times [-h, \eta],$$

$$\eta_t + \varphi_{(x)}\eta_x + \varphi_{(y)}\eta_y - \varphi_{(z)} = 0,$$

$$\varphi_{(t)} + \frac{1}{2}\left(\varphi_{(x)}^2 + \varphi_{(y)}^2 + \varphi_{(z)}^2\right) + g\eta - \frac{\sigma}{\rho}C = 0,$$

$$\phi_z = 0 \quad \text{on } z = -h.$$

The conservation of momentum equation (2.16) is not stated; it is used to find the pressure p once η and ϕ have been found. In water of infinite depth, the kinematic boundary condition on the bottom (last equation) is replaced by

$$|\nabla\phi| \to 0 \quad \text{as } z \to -\infty.$$

The surface wave problem has been studied for more than a century. It is a difficult nonlinear problem because of its two nonlinear boundary conditions on the free surface. Assumptions can be made to simplify the problem : linearization of the equations, restriction to periodic solutions or to steady solutions in a moving frame of reference, flat bottom, two spatial dimensions only, model equations.

Linearization of the 2D problem and dispersion relation One looks for solutions which are periodic in x with wave number k and in t with frequency ω. The following dimensionless variables are introduced :

$$(x^*, z^*) = (kx, kz), \quad \eta^* = \frac{\eta}{a}, \quad t^* = \omega t, \quad \phi^* = \frac{k}{\omega a}\phi,$$

where a denotes the amplitude of the wave. In terms of the dimensionless variables and after dropping the stars, the surface wave problem linearized around the equilibrium state $\eta = 0$, $\mathbf{u} = \mathbf{0}$ reads :

$$\phi_{xx} + \phi_{zz} = 0 \quad \text{for } (x, z) \in \mathbb{R} \times [-kh, 0],$$

$$\eta_t - \phi_z = 0 \quad \text{on } z = 0,$$

$$\phi_t + \left(\frac{gk}{\omega^2}\right)\eta - \left(\frac{\sigma k^3}{\rho\omega^2}\right)\eta_{xx} = 0 \quad \text{on } z = 0,$$

$$\phi_z = 0 \quad \text{on } z = -kh.$$

Consider solutions in the form of sinusoidal dispersive waves $\eta(x, t) = \cos(x - t + \Theta)$. In the linear theory, the wave number k and the frequency ω are assumed to be constant. Since the equations are linear, the amplitude is arbitrary. On the other hand, for the boundary conditions on the free surface to be satisfied, ω and k must satisfy a *dispersion relation* of the form $D(\omega, k) = 0$. Note that dimensional analysis immediately gives

$$\frac{\omega}{\sqrt{gk}} = \mathcal{G}\left(kh, \frac{\sigma k^2}{\rho g}, ka\right).$$

Let us now obtain the function $\mathcal{G}$. Laplace's equation combined with the kinematic conditions on the bottom and on the free surface gives

$$\phi(x - t, z) = \frac{1}{\tanh(kh)} \sin(x - t + \Theta) \frac{\cosh(z + kh)}{\cosh(kh)} .$$

The dynamic condition on the free surface yields

$$D(\omega, k) \equiv \omega^2 - gk \tanh(kh) \left(1 + \frac{\sigma k^2}{\rho g}\right) = 0 . \tag{2.23}$$

The final expressions for the potential ϕ and for the elevation of the free surface η in the original variables are

$$\phi = b \sin(kx - \omega t + \Theta) \frac{\cosh(kz + kh)}{\cosh(kh)} , \quad \eta = a \cos(kx - \omega t + \Theta) , \tag{2.24}$$

where

$$b = a \frac{g}{\omega} \left(1 + \frac{\sigma k^2}{\rho g}\right) .$$

Equation (2.23) is the well-known dispersion relation for linearized 2D capillary–gravity periodic waves in water of finite depth. Written in terms of the phase speed c rather than the frequency ω, the dispersion relation becomes

$$F^2 = \frac{\tanh(kh)}{kh} \left(1 + (kh)^2 W\right) , \tag{2.25}$$

where the Froude number F and the Weber number W are defined as

$$F = \frac{c}{\sqrt{gh}} , \quad W = \frac{\sigma}{\rho g h^2} .$$

Equation (2.25) has been plotted in Figure 1 for three different values of W. When $0 < W < 1/3$, the curve has, in addition to the extremum at the origin, a minimum where phase and group velocities are equal. Note that in infinite depth the dispersion relation becomes

$$D(\omega, k) \equiv \omega^2 - g|k| \left(1 + \frac{\sigma k^2}{\rho g}\right) = 0 . \tag{2.26}$$

The presence of the absolute value has subtle consequences.

For the air/water interface, the minimum is obtained (in infinite depth) for a wave length of 1.73 cm and a wave speed of 23.2 cm/s. For short waves (kh large), surface tension has a stronger effect than gravity. For long waves (kh small), it is the opposite. If the Froude number F is between the minimum value and one (and $0 < W < 1/3$), there are two corresponding positive wave numbers, one on the gravity side and one on the capillary side.

In water of 10 centimeter depth, the critical speed $\sqrt{gh}$ is 1m/s. For the air/water interface, the corresponding short wave having the same phase velocity is given by (since kh is large)

$$kh \approx \frac{\sigma}{\rho g h^2} (kh)^2 = 0.00074 (kh)^2, \quad \text{or} \quad kh \approx 1350 ,$$

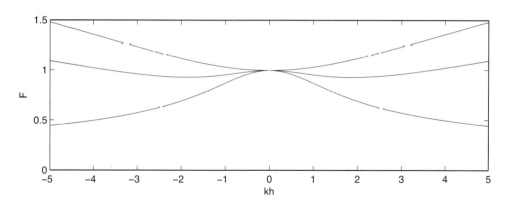

Figure 1. Dispersion relation (2.25) for capillary–gravity waves. The three curves correspond, from top to bottom, to $W = 0.4$, $W = 0.2$ and $W = 0$.

i.e. its wavelength is half a millimeter. If $h = 5$ cm, the wavelength is 1 mm.

The group velocity can be written as

$$c_g = c \left[\frac{1}{2} \left(\frac{1 + 3(kh)^2 W}{1 + (kh)^2 W} \right) + \frac{kh}{\sinh 2kh} \right].$$

For long waves ($kh \to 0$), the group and phase velocities are equal to $\sqrt{gh}$. There is no dispersion. For pure capillary waves ($g = 0$), the group velocity is

$$c_g = c \left(\frac{3}{2} + \frac{kh}{\sinh 2kh} \right),$$

while, for pure gravity waves ($\sigma = 0$), the group velocity is

$$c_g = c \left(\frac{1}{2} + \frac{kh}{\sinh 2kh} \right).$$

A first interpretation of the group velocity is given by considering the superposition of two linear waves, with almost equal wave numbers and frequencies :

$$\begin{aligned} \eta_1 &= A\cos(kx - \omega t), \\ \eta_2 &= A\cos[(k + \delta k)x - (\omega + \delta\omega)t]. \end{aligned}$$

The resulting profile is given by

$$\begin{aligned} \eta = \eta_1 + \eta_2 &= 2A \cos\left[\tfrac{1}{2}(\delta k\, x - \delta\omega\, t)\right] \cos\left[kx - \omega t + \tfrac{1}{2}(\delta k\, x - \delta\omega\, t)\right] \\ &\approx 2A \cos\left[\tfrac{1}{2}(\delta k\, x - \delta\omega\, t)\right] \cos(kx - \omega t) \end{aligned} \qquad (2.27)$$

and is shown in Figure 2. The envelope (first cosine term) travels at the group velocity

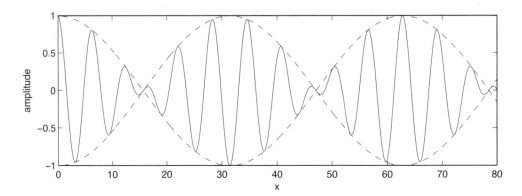

Figure 2. Wave packet (2.27) at time $t = 0$ with $k = 1$, $\delta k = 0.2$ and $A = 1/2$. The envelope is shown in dashed lines.

while the carrying wave inside the envelope second cosine term) travels at the phase velocity.

A second interpretation of the group velocity is given by considering the speed of propagation of energy. We define the kinetic energy density K and the average kinetic energy over a wave length $\overline{K}$

$$K = \int_{-h}^{\eta} \tfrac{1}{2}\rho|\nabla\phi|^2 \, dz, \quad \overline{K} = \frac{1}{L}\int_{x}^{x+L} K \, dx, \tag{2.28}$$

as well as the potential energy density V and the average potential energy over a wave length $\overline{V}$

$$V = \tfrac{1}{2}\rho g\eta^2 + \sigma\left(\sqrt{1 + \eta_x^2} - 1\right), \quad \overline{V} = \frac{1}{L}\int_{x}^{x+L} V \, dx, \tag{2.29}$$

where L denotes the wave length. The total energy density is $E = K + V$.

For linearized waves, the expressions for $\overline{K}$ and $\overline{V}$ become

$$\overline{K} = \frac{1}{L}\int_{x}^{x+L}\int_{-h}^{0} \frac{1}{2}\rho|\nabla\phi|^2 \, dz \, dx = \frac{1}{4}\rho g a^2\left(1 + \frac{\sigma k^2}{\rho g}\right),$$

$$\overline{V} = \frac{1}{L}\int_{x}^{x+L}\left(\frac{1}{2}\rho g\eta^2 + \frac{1}{2}\sigma\eta_x^2\right) dx = \frac{1}{4}\rho g a^2\left(1 + \frac{\sigma k^2}{\rho g}\right).$$

Potential and kinetic energy are equal. This result, however, is no longer true for nonlinear waves.

For the sake of simplicity, let us now neglect the effects due to surface tension. The kinetic energy balance for the flow of a perfect fluid can be written as

$$\frac{\partial}{\partial t}\int_{\Omega} \tfrac{1}{2}\rho|\nabla\phi|^2 \, d\Omega + \int_{\partial\Omega} \tfrac{1}{2}\rho|\nabla\phi|^2 \, \mathbf{u}\cdot\mathbf{n} \, da = \int_{\Omega} \rho\, \mathbf{f}\cdot\mathbf{u} \, d\Omega - \int_{\partial\Omega} p\, \mathbf{u}\cdot\mathbf{n} \, da,$$

where $\mathbf{f}$ is the force due to gravity, Ω the domain contained in the rectangle between two vertical lines at $x = x_1$ and at $x = x_2$, the bottom and a horizontal line above the free surface, $\partial\Omega$ the boundary of this domain.

Let us transform first the term containing the gravity force, by introducing the potential $\Phi = -gz$. The free surface is denoted by Σ.

$$
\begin{aligned}
\int_\Omega \rho\, \mathbf{f} \cdot \mathbf{u}\, d\Omega &= \int_\Omega \rho\, \nabla\Phi \cdot \mathbf{u}\, d\Omega \\
&= \int_\Omega \rho\, \nabla \cdot (\Phi\mathbf{u})\, d\Omega \quad (\text{since } \nabla \cdot \mathbf{u} = 0) \\
&= \int_{\partial\Omega} \rho\, \Phi\, \mathbf{u} \cdot \mathbf{n}\, da - \int_\Sigma \rho\, [[\Phi\mathbf{u}]] \cdot \mathbf{n}\, d\Sigma \\
&\quad (\text{divergence theorem in the presence of a free surface}) \\
&= -\int_{\partial\Omega} \rho g z\, \mathbf{u} \cdot \mathbf{n}\, da - \int_\Sigma \frac{\rho g \eta \eta_t}{\sqrt{1 + \eta_x^2}}\, d\Sigma\,. \\
&\quad (\text{kinematic condition})
\end{aligned}
$$

Therefore the right hand side of the kinetic energy balance can be rewritten as

$$
\begin{aligned}
\int_\Omega \rho\, \mathbf{f} \cdot \mathbf{u}\, d\Omega - \int_{\partial\Omega} p\, \mathbf{u} \cdot \mathbf{n}\, da &= -\int_{\partial\Omega} (p + \rho g z)\, \mathbf{u} \cdot \mathbf{n}\, da - \int_\Sigma \frac{\rho g \eta \eta_t}{\sqrt{1 + \eta_x^2}}\, d\Sigma \\
&= \int_{\partial\Omega} \rho \left(\frac{\partial\phi}{\partial t} + \tfrac{1}{2}|\nabla\phi|^2 \right) \mathbf{u} \cdot \mathbf{n}\, da - \int_\Sigma \frac{\rho g \eta \eta_t}{\sqrt{1 + \eta_x^2}}\, d\Sigma\,.
\end{aligned}
$$

It follows that

$$
\frac{\partial}{\partial t} \int_\Omega \tfrac{1}{2}\rho\, |\nabla\phi|^2\, d\Omega + \int_\Sigma \frac{\rho g \eta \eta_t}{\sqrt{1 + \eta_x^2}}\, d\Sigma = \int_{\partial\Omega} \rho\, \frac{\partial\phi}{\partial t}\, \mathbf{u} \cdot \mathbf{n}\, da\,,
$$

or $\displaystyle \frac{\partial}{\partial t} \int_{x_1}^{x_2} \left(\int_{-h}^{\eta} \tfrac{1}{2}\rho\, |\nabla\phi|^2\, dz + \tfrac{1}{2}\rho g \eta^2 \right) dx = -\mathcal{F}(x_2) + \mathcal{F}(x_1)\,.$

It yields the equation

$$
\frac{\partial E}{\partial t} + \frac{\partial \mathcal{F}}{\partial x} = 0\,,
$$

where $\mathcal{F}$ is the energy flux defined by

$$
\mathcal{F} = -\int_{-h}^{\eta} \rho \phi_x \phi_t\, dz\,.
$$

A local derivation of that equation can also be given.

For the waves (2.24) of the linearized problem with $\Theta = 0$, one computes easily

$$
E = \frac{1}{2}\rho g a^2 \cos^2(kx - \omega t) \left(1 + \frac{2kh}{\sinh(2kh)} \right) + \frac{1}{2}\rho g a^2 \left(\frac{1}{2} - \frac{kh}{\sinh(2kh)} \right),
$$

and

$$\mathcal{F} = \frac{1}{2}\rho g a^2 \frac{\omega}{k} \cos^2(kx - \omega t)\left(1 + \frac{2kh}{\sinh(2kh)}\right).$$

By taking the average of these two expressions over a period, one finds

$$\overline{E} = \frac{1}{2}\rho g a^2 \quad \text{and} \quad \overline{\mathcal{F}} = \frac{1}{2}\rho g c_g a^2.$$

It follows that $\overline{\mathcal{F}} = c_g \overline{E}$. The energy propagates with the group velocity.

Illustrations of the concept of group velocity   There are several experiments which clearly illustrate the concept of group velocity. We have seen that there are two interpretations of the concept of group velocity, one which is purely kinematic and one which is purely dynamical. The link between the two is not that clear (see Stoker (1958)). Stoker gives a preference to the kinematic interpretation. Three well-known illustrations of the concept of group velocity are the wave-maker, the fish-line problem and the ship wake. The main feature of these three examples is that the free surface is perturbed by a surface piercing object. In the problem of the wave maker, a paddle oscillates and disturbs the free surface. In the problem of the fish-line, the object is fixed but the flow has a current. In the problem of the ship wake, the ship is moving on the free surface of a fluid otherwise at rest.

Dispersion relation for 3D waves   The extension of the dispersion relation (2.23) to three-dimensional waves is

$$D(\omega, k, l) \equiv \omega^2 - g|\mathbf{k}|\tanh(|\mathbf{k}|h)\left(1 + \frac{\sigma|\mathbf{k}|^2}{\rho g}\right) = 0, \tag{2.30}$$

where $|\mathbf{k}| = \sqrt{k^2 + l^2}$.

A brief account of the method of multiple scales to derive the NLS equation from the full water-wave problem   Accounting for nonlinear and dispersive effects correct to third order in the wave steepness, the envelope of a weakly nonlinear gravity–capillary wavepacket in deep water is governed by the NLS equation. A more accurate envelope equation, which includes effects up to fourth order in the wave steepness, was derived by Dysthe (1979) for pure gravity wavepackets. He used the method of multiple scales. Later, Stiassnie (1984) showed that the Dysthe equation is merely a particular case of the more general Zakharov equation that is free of the narrow spectral width assumption. Hogan (1985) extended Stiassnie's results to deep-water gravity–capillary wavepackets. Apart from the leading-order nonlinear and dispersive terms present in the NLS equation, the fourth-order equation of Hogan features certain nonlinear modulation terms and a nonlocal term that describes the coupling of the envelope with the induced mean flow. In addition to playing a significant part in the stability of a uniform wavetrain, this mean flow turns out to be important at the tails of gravity–capillary solitary waves in deep water as shown by Akylas, Dias and Grimshaw (1998).

The common ansatz used in the derivation of the NLS equation (or of the Dysthe equation if one goes to higher order) is that the velocity potential ϕ and the free-surface elevation η have uniformly valid asymptotic expansions in terms of a small parameter ϵ (the dimensionless amplitude of the wave, kA, for example). One writes

$$\eta = \sum_{n=1}^{3} \epsilon^n \, \eta_n(x_0, x_1, x_2, y_1, y_2; t_0, t_1, t_2) + \mathcal{O}(\epsilon^4), \tag{2.31}$$

$$\phi = \sum_{n=1}^{3} \epsilon^n \, \phi_n(x_0, x_1, x_2, y_1, y_2, z; t_0, t_1, t_2) + \mathcal{O}(\epsilon^4), \tag{2.32}$$

where

$$x_0 = x \, , x_1 = \epsilon \, x \, , x_2 = \epsilon^2 \, x \, , y_1 = \epsilon \, y \, , y_2 = \epsilon^2 \, y \, , t_0 = t \, , t_1 = \epsilon \, t \, , t_2 = \epsilon^2 \, t . \tag{2.33}$$

The order one component of η is the linearized monochromatic wave

$$\eta_1 = A \, e^{i(kx - \omega t)} + \text{c.c.} . \tag{2.34}$$

Applying the method of multiple scales leads to the following equation for the evolution of the complex amplitude A of the wave :

$$2i \frac{\partial A}{\partial t_2} + p \frac{\partial^2 A}{\partial \xi^2} + q \frac{\partial^2 A}{\partial y_1^2} + \gamma A |A|^2 = -i\epsilon \left(s A_{\xi y_1 y_1} + r A_{\xi\xi\xi} + u A^2 A^*_\xi - v |A|^2 A_\xi \right)$$
$$+ \epsilon A \overline{\phi}_\xi \big|_{z_1 = 0} . \tag{2.35}$$

The evolution equation has been written in dimensionless form, with all lengths nondimensionalized by k, time by ω and potential by $2k^2/\omega$. Here $\xi = x_1 - c_g t_1$ and $z_1 = \epsilon z$ are scaled variables that describe the wavepacket modulations in a frame of reference moving with the group velocity c_g. As expected, to leading order in the wave steepness $\epsilon \ll 1$, equation (2.35) reduces to the familiar NLS equation, while the coupling with the induced mean flow is reflected in the last term of (2.35). Specifically, the mean-flow velocity potential $\epsilon^2 \overline{\phi}(\xi, z_1, t_2)$ satisfies the boundary-value problem

$$\overline{\phi}_{\xi\xi} + \overline{\phi}_{z_1 z_1} = 0 \qquad (-\infty < z_1 < 0, \ -\infty < \xi < \infty) ,$$
$$\overline{\phi}_{z_1} = (|A|^2)_\xi \qquad (z_1 = 0) ,$$
$$\overline{\phi} \rightarrow 0 \qquad (z_1 \rightarrow -\infty) ,$$

from which it follows that

$$\overline{\phi}_\xi \big|_{z_1 = 0} = - \int_{-\infty}^{\infty} |s| \, e^{is\xi} \, FT(|A|^2) \, ds , \tag{2.36}$$

where

$$FT(\cdot) = \frac{1}{2\pi} \int_{-\infty}^{\infty} e^{-is\xi} \, (\cdot) \, d\xi$$

denotes the Fourier transform. Hence, the coupling of the envelope with the induced mean flow enters via a nonlocal term in the fourth-order envelope equation. The coefficients of the rest of the terms in (2.35) are given by the following expressions, where $B = \sigma k^2/\rho g$:

$$p = \frac{k^2}{\omega}\frac{d^2\omega}{dk^2} = \frac{3B^2 + 6B - 1}{4(1+B)^2}, \tag{2.37}$$

$$q = \frac{k}{\omega}\frac{d\omega}{dk} = \frac{1 + 3B}{2(1+B)}, \tag{2.38}$$

$$\gamma = -\frac{2B^2 + B + 8}{8(1 - 2B)(1+B)}, \tag{2.39}$$

$$r = -\frac{(1 - B)(B^2 + 6B + 1)}{8(1+B)^3}, \tag{2.40}$$

$$s = \frac{3 + 2B + 3B^2}{4(1+B)^2}, \tag{2.41}$$

$$u = \frac{(1 - B)(2B^2 + B + 8)}{16(1 - 2B)(1+B)^2}, \tag{2.42}$$

$$v = \frac{3(4B^4 + 4B^3 - 9B^2 + B - 8)}{8(1 - 2B)^2(1+B)^2}. \tag{2.43}$$

These expressions are not valid when B is close to $1/2$, which corresponds to the resonance between the first and second harmonics.

Derivation from Zakharov's integral equation   This derivation was first considered by Stiassnie (1984) for gravity waves and by Hogan (1985) for capillary–gravity waves. Instead of working directly with the free-surface elevation $\eta(\mathbf{x}, t)$, the new variable $B(\mathbf{k}, t)$ is introduced. The link between η and B is :

$$\eta(\mathbf{x}, t) = \frac{1}{2\pi}\int_{-\infty}^{+\infty}\left(\frac{|\mathbf{k}|}{2\omega(\mathbf{k})}\right)^{1/2}\{B(\mathbf{k}, t)\exp\{i[\mathbf{k}\cdot\mathbf{x} - \omega(\mathbf{k})t]\} + \text{c.c.}\}\,d\mathbf{k}. \tag{2.44}$$

Zakharov's integral equation for $B(\mathbf{k}, t)$ is (see also Chapter 5)

$$i\frac{\partial B}{\partial t}(\mathbf{k}, t) = \iiint_{-\infty}^{+\infty}T(\mathbf{k}, \mathbf{k}_1, \mathbf{k}_2, \mathbf{k}_3)B^*(\mathbf{k}_1, t)B(\mathbf{k}_2, t)B(\mathbf{k}_3, t)\delta(\mathbf{k} + \mathbf{k}_1 - \mathbf{k}_2 - \mathbf{k}_3)$$

$$\times \exp\{i[\omega(\mathbf{k}) + \omega(\mathbf{k}_1) - \omega(\mathbf{k}_2) - \omega(\mathbf{k}_3)]t\}\,d\mathbf{k}_1\,d\mathbf{k}_2\,d\mathbf{k}_3, \tag{2.45}$$

where the wave vector $\mathbf{k}$ and the frequency ω are related through the linear dispersion relation (2.30)

$$\omega(\mathbf{k}) = \left(g|\mathbf{k}| + \sigma/\rho\,|\mathbf{k}|^3\right)^{1/2}.$$

The function $T(\mathbf{k}, \mathbf{k}_1, \mathbf{k}_2, \mathbf{k}_3)$ is a lengthy scalar function given for example by Krasitskii (1990). Equation (2.45) describes four-wave interaction processes obeying the resonant conditions

$$\mathbf{k} + \mathbf{k}_1 = \mathbf{k}_2 + \mathbf{k}_3, \tag{2.46}$$

$$\omega(\mathbf{k}) + \omega(\mathbf{k}_1) = \omega(\mathbf{k}_2) + \omega(\mathbf{k}_3). \tag{2.47}$$

But, as pointed out by Zakharov (1968), there are difficulties in applying equation (2.45) to capillary–gravity waves. This is because, unlike gravity waves, these waves can satisfy triad resonances. The condition for triad resonance will give a zero denominator in one of the terms of $T(\mathbf{k}, \mathbf{k}_1, \mathbf{k}_2, \mathbf{k}_3)$ corresponding to the second-order interaction. However, if the wave packet is sufficiently narrow, then the resonance condition cannot be satisfied.

Let $\mathbf{k} = \mathbf{k}_0 + \chi$ where $\mathbf{k}_0 = (k, 0)$ and $\chi = (p, q)$. Let ϵ denote the order of the spectral width $|\chi|/k$. Let also $\omega(\mathbf{k}_0) = \omega$ and $\chi_i = (p_i, q_i), i = 1, 2, 3$.

Introducing a new variable $A(\chi, t)$ given by

$$A(\chi, t) = B(\mathbf{k}, t) \exp\{-i[\omega(\mathbf{k}) - \omega(\mathbf{k}_0)]t\} \tag{2.48}$$

in equations (2.45) and (2.44), one obtains

$$i\frac{\partial A}{\partial t}(\chi, t) - [\omega(\mathbf{k}) - \omega]A(\chi, t) = \iiint_{-\infty}^{+\infty} T(\mathbf{k}_0 + \chi, \mathbf{k}_0 + \chi_1, \mathbf{k}_0 + \chi_2, \mathbf{k}_0 + \chi_3) \tag{2.49}$$
$$\delta(\chi + \chi_1 - \chi_2 - \chi_3)A^*(\chi_1)A(\chi_2)A(\chi_3)\, d\chi_1\, d\chi_2\, d\chi_3 ,$$

and

$$\eta(\mathbf{x}, t) = \exp\{i(kx - \omega t)\}\frac{1}{2\pi}\int_{-\infty}^{+\infty}\left(\frac{|\mathbf{k}|}{2\omega(\mathbf{k})}\right)^{1/2} A(\chi, t)\, e^{i\chi \cdot \mathbf{x}}\, d\chi + \text{c.c.} . \tag{2.50}$$

The Taylor expansion of $|\mathbf{k}_0 + \chi|/2\omega(\mathbf{k}_0 + \chi)$ in powers of $|\chi|/k$ is

$$\left(\frac{|\mathbf{k}_0 + \chi|}{2\omega(\mathbf{k}_0 + \chi)}\right)^{1/2} = \left(\frac{\omega}{2g(1 + B)}\right)^{1/2}\left(1 + \frac{p}{4k}\left(\frac{1 - B}{1 + B}\right)\right), \tag{2.51}$$

in which terms up to order ϵ have been retained.

Substituting equation (2.51) into (2.50), $\eta(\mathbf{x}, t)$ can be expressed as

$$\eta(\mathbf{x}, t) = \text{Re}\left\{a(\mathbf{x}, t)e^{i(kx - \omega t)}\right\}, \tag{2.52}$$

where

$$a(\mathbf{x}, t) = \frac{1}{2\pi}\left(\frac{2\omega}{g(1 + B)}\right)^{1/2}\int_{-\infty}^{+\infty}\left(1 + \frac{p}{4k}\left(\frac{1 - B}{1 + B}\right)\right) A(\chi, t)e^{i\chi \cdot \mathbf{x}}\, d\chi . \tag{2.53}$$

By Taylor expanding $\omega(\mathbf{k}) - \omega$ in powers of $|\chi|/k$ and keeping terms up to order ϵ we get

$$\omega(\mathbf{k}) - \omega = \frac{1}{2}\left(\frac{g}{k(1 + B)}\right)^{1/2}\left\{p(1 + 3B) + \frac{p^2}{4k}\left(\frac{-1 + 6B + 3B^2}{1 + B}\right) + \frac{q^2}{2k}(1 + 3B)\right.$$
$$\left. + \frac{p^3}{8k^2}\left(\frac{(1 - B)(1 + 6B + B^2)}{(1 + B)^2}\right) - \frac{pq^2}{4k^2}\left(\frac{3 + 2B + 3B^2}{1 + B}\right)\right\}. \tag{2.54}$$

In (2.49), we substitute the expression (2.54) for $\omega(\mathbf{k}) - \omega$. By replacing $A(\chi, t)$ by $a(\mathbf{x}, t)$ and taking the inverse Fourier transform, one finds

$$i\, a_t + \frac{1}{2}\left(\frac{g}{k(1 + B)}\right)^{1/2}\left\{i(1 + 3B)a_x + \left(\frac{-1 + 6B + 3B^2}{4k(1 + B)}\right)a_{xx} + \frac{(1 + 3B)}{2k}a_{yy}\right.$$
$$\left. -i\frac{(1 - B)(1 + 6B + B^2)}{8k^2(1 + B)^2}a_{xxx} + i\frac{3 + 2B + 3B^2}{4k^2(1 + B)}a_{xyy}\right\}$$

$$= \frac{1}{2\pi}\left(\frac{2\omega}{g(1+B)}\right)^{1/2}\iiint_{-\infty}^{+\infty}\left[1+\frac{p_2+p_3-p_1}{4k}\left(\frac{1-B}{1+B}\right)\right]$$
$$\times T(\mathbf{k}_0+\chi_2+\chi_3-\chi_1, \mathbf{k}_0+\chi_1, \mathbf{k}_0+\chi_2, \mathbf{k}_0+\chi_3)$$
$$\times A^*(\chi_1)A(\chi_2)A(\chi_3)\exp\{i(\chi_2+\chi_3-\chi_1)\cdot\mathbf{x}\}\,d\chi_1\,d\chi_2\,d\chi_3\,. \tag{2.55}$$

Now it can be shown that the Taylor expansion of T keeping terms up to order ϵ becomes

$$T(\mathbf{k}_0+\chi_2+\chi_3-\chi_1, \mathbf{k}_0+\chi_1, \mathbf{k}_0+\chi_2, \mathbf{k}_0+\chi_3) = \frac{k^3}{8\pi^2}\left[\frac{(8+B+2B^2)}{4(1+B)(1-2B)}\right.$$

$$\left.+\frac{3(p_2+p_3)}{8k}\frac{(8-B+9B^2-4B^3-4B^4)}{(1+B)^2(1-2B)^2}-\frac{(p_3-p_1)^2}{k|\chi_1-\chi_3|}-\frac{(p_2-p_1)^2}{k|\chi_1-\chi_2|}\right]. \tag{2.56}$$

By using this form of T, we find by using (2.53) that the right-hand side of (2.55) becomes on integration

$$\frac{g}{16\omega}\left[\frac{k^3(8+B+2B^2)}{(1-2B)}a|a|^2 - \tfrac{1}{2}i\,k^2\frac{(1-B)(8+B+2B^2)}{(1+B)(1-2B)}a^2a_x^*\right.$$

$$\left.-3i\,k^2\frac{(8-B+9B^2-4B^3-4B^4)}{(1+B)(1-2B)^2}|a|^2a_x\right] - \frac{k^2}{4\pi^2}aI\,, \tag{2.57}$$

where

$$I = \iint_{-\infty}^{+\infty}\frac{(p_1-p_2)^2}{|\chi_1-\chi_2|}A^*(\chi_1)A(\chi_2)e^{i(\chi_2-\chi_1)\cdot\mathbf{x}}\,d\chi_1\,d\chi_2\,. \tag{2.58}$$

It can be shown that

$$I = \left[\frac{g(1+B)}{2\omega}\right]2\pi\int_{-\infty}^{+\infty}\frac{\partial}{\partial X}(|a|^2)\frac{x-X}{|\mathbf{x}-\mathbf{X}|^3}\,d\mathbf{X}\,. \tag{2.59}$$

The integral I can be related to the mean-flow velocity potential $\overline{\phi}$ (2.36). Collecting the results from equations (2.55) and (2.59), together with those from the expression (2.57), and making the same scaling transformation as above leads to equation (2.35). More details can be found in Dias and Kharif (1999). See also the interesting paper by Phillips (1981), which provides a history of wave interactions.

3 Mathematical properties

In this section, we consider the NLS equation in the following form :

$$i\,u_t + \alpha\,u_{xx} + \gamma\,|u|^2u = 0\,, \quad u \in \mathbb{C}. \tag{3.1}$$

This form can always been obtained with rescaling. We emphasize here that there is only one NLS equation (focussing if $\alpha\gamma > 0$ and defocussing if $\alpha\gamma < 0$) for all physical systems. Incidentally the NLS equation (2.14) derived from the KdV equation is of the defocussing type. The NLS equation derived from the water-wave problem is of the focussing type if $kh > 1.363$ and of the defocussing type if $kh < 1.363$.

3.1   Hamiltonian structure

The NLS equation (3.1) is a Hamiltonian partial differential equation. Taking real coordinates, the following Hamiltonian system can be derived. Let $u = u_1 + iu_2$ and write $\mathbf{U} = (u_1, u_2)$. Then the NLS equation takes the form

$$JU_t + \alpha \mathbf{U}_{xx} + \gamma \|\mathbf{U}\|^2 \mathbf{U} = 0 \quad \text{where} \quad J = \begin{pmatrix} 0 & -1 \\ 1 & 0 \end{pmatrix}, \tag{3.2}$$

or

$$J\mathbf{U}_t = \nabla H(\mathbf{U}), \tag{3.3}$$

with

$$H(\mathbf{U}) = \int_{\mathbb{R}} \left[\tfrac{1}{2}\alpha \|\mathbf{U}_x\|^2 - \tfrac{1}{4}\gamma \|\mathbf{U}\|^4 \right] \, dx. \tag{3.4}$$

The expression for H clearly shows why the case $\alpha\gamma > 0$ is called focussing.

3.2   Conservation laws

There is an infinite number of conservation laws satisfied by solutions to the NLS equation. Only two will be of concern to us. The first one is the conservation law for $|u(x,t)|^2$, which is given by

$$\frac{\partial}{\partial t}|u|^2 = i\alpha \frac{\partial}{\partial x}(\overline{u}u_x - u\overline{u}_x).$$

Suppose $|u|^2$ can be integrated over an x–interval in such a way that $(\overline{u}u_x - u\overline{u}_x)$ vanishes at the boundary of the interval. Then

$$\frac{d}{dt}\int |u|^2 \, dx = 0.$$

The second one is the energy conservation. Define the energy density $E(x,t)$ and the energy flux $\mathcal{F}(x,t)$ by

$$\begin{aligned} E(x,t) &= \tfrac{1}{2}\alpha |u_x|^2 - \tfrac{1}{4}\gamma |u|^4 & (3.5) \\ \mathcal{F}(x,t) &= -\tfrac{1}{2}\alpha(\overline{u_x}u_t + u_x\overline{u_t}). & (3.6) \end{aligned}$$

Note that the energy is indefinite : the integral of $E(x,t)$ over x can be negative or positive, depending on the initial data. A calculation shows that the energy conservation law is

$$\frac{\partial E}{\partial t} + \frac{\partial \mathcal{F}}{\partial x} = 0. \tag{3.7}$$

Integrating over x and assuming that the boundary conditions are such that $\int \partial \mathcal{F}/\partial x \, dx = 0$ yields

$$\frac{d}{dt}\int \left(\tfrac{1}{2}\alpha |u_x|^2 - \tfrac{1}{4}\gamma |u|^4 \right) \, dx = 0.$$

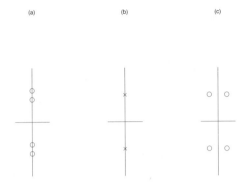

Figure 3. Schematic of the eigenvalue movement associated with the Benjamin–Feir instability. The circles indicate simple eigenvalues, while the crosses indicate double eigenvalues.

4 An NLS model for the Benjamin–Feir instability

The discovery of the Benjamin–Feir (BF) instability of Stokes travelling waves was a milestone in the history of water waves. It is due to the seminal work of Benjamin and Feir (1967) which combined experimental evidence with a weakly nonlinear theory. Mathematically, the BF instability can be characterized as a collision of two pairs of purely imaginary eigenvalues of opposite energy sign (see Figure 3 for a schematic). For the full water-wave problem, there is in addition to the four eigenvalues shown in Figure 3 a countable number of stable purely imaginary eigenvalues. This point of view of the BF instability was a byproduct of the proof of the BF instability provided by Bridges and Mielke (1995). It also appears in the nonlinear Schrödinger model, where it can be shown explicitly. One of the pairs of modes in the collision has negative energy (relative to the basic state) and one pair has positive energy.

As said above, the NLS equation describes the modulations of weakly nonlinear, deep-water gravity waves, with basic wave number k_0 and frequency $\omega_0(k_0)$:

$$\mathrm{i}\,(A_T + c_g A_X) - \frac{\omega_0}{8k_0^2}A_{XX} - \frac{1}{2}\omega_0 k_0^2 |A|^2 A = 0, \tag{4.1}$$

where A varies slowly in X and T, and $c_g = \omega_0/(2k_0)$ is the group velocity. At leading order, the free-surface elevation $\eta(x,t)$ is given by

$$\eta(x,t) = A(X,T)e^{i(k_0 x - \omega_0 t)} + \text{c.c.}\,. \tag{4.2}$$

In a frame of reference moving with the group velocity, equation (4.1) becomes

$$\mathrm{i}\,A_T - \frac{\omega_0}{8k_0^2}A_{\xi\xi} - \frac{1}{2}\omega_0 k_0^2 |A|^2 A = 0, \quad \text{with} \quad \xi = X - c_g T. \tag{4.3}$$

By scaling T, ξ and changing $\overline{A}$ into u, one obtains the normalized equation (3.1), where α and γ are taken to be positive.

The basic travelling wave solution   Equation (3.1) admits travelling wave solutions. Let $\theta(x,t) - kx + \omega t + \Theta$, and consider the basic travelling wave solution to (3.2)

$$\widehat{\mathbf{U}}(x,t) = R_{\theta(x,t)}\mathbf{U}_0, \quad R_\theta = \begin{pmatrix} \cos\theta & -\sin\theta \\ \sin\theta & \cos\theta \end{pmatrix}. \tag{4.4}$$

Then $\mathbf{U}_0, \omega, k$ satisfy

$$\omega + \alpha k^2 = \gamma \|\mathbf{U}_0\|^2. \tag{4.5}$$

Below we show that the condition for a BF instability of the basic travelling wave solution of (3.1) is $\alpha\gamma > 0$, a condition satisfied by deep water waves. The BF instability does not occur in the defocussing NLS equation ($\alpha\gamma < 0$).

Formulating the conservative BF stability problem   Linearize the partial differential equation (3.2) about the basic travelling wave (4.4). Let $\mathbf{U}(x,t) = R_{\theta(x,t)}(\mathbf{U}_0 + \mathbf{V}(x,t))$, substitute into the governing equation (3.2), linearize about $\mathbf{U}_0$, and simplify using (4.5). One gets

$$J\mathbf{V}_t + 2\alpha k J\mathbf{V}_x + \alpha \mathbf{V}_{xx} + 2\gamma\langle \mathbf{U}_0, \mathbf{V}\rangle \mathbf{U}_0 = \mathbf{0}, \tag{4.6}$$

where $\langle\cdot,\cdot\rangle$ is the standard scalar product on $\mathbb{R}^2$.

The class of solutions of interest are solutions which are periodic in x with wavenumber σ. The parameter σ represents the sideband. The BF instability will be associated with the limit $|\sigma| \ll 1$. Therefore let

$$\mathbf{V}(x,t) = \tfrac{1}{2}\mathbf{V}_0(t) + \sum_{n=1}^{\infty}(\mathbf{V}_n(t)\cos n\sigma x + \mathbf{W}_n(t)\sin n\sigma x).$$

The $\sigma-$independent mode satisfies

$$J\partial_t\mathbf{V}_0 + 2\gamma\mathbf{U}_0\mathbf{U}_0^T\mathbf{V}_0 = \mathbf{0}. \tag{4.7}$$

Letting $\mathbf{V}_0(t) = e^{\lambda t}\widetilde{\mathbf{V}}_0$, it is easy to verify that λ satisfies $\lambda^2 = 0$. All solutions of this equation are neutral, and therefore do not contribute to instability. The $\sigma-$independent modes are associated with the *superharmonic* instability which is known to be stable at low amplitude.

The $\sigma-$dependent modes decouple into 4$-$dimensional subspaces for each n, and satisfy

$$\begin{aligned} J\partial_t\mathbf{V}_n + 2\alpha kn\sigma J\mathbf{W}_n - \alpha(n\sigma)^2\mathbf{V}_n + 2\gamma\mathbf{U}_0\mathbf{U}_0^T\mathbf{V}_n &= \mathbf{0} \\ J\partial_t\mathbf{W}_n - 2\alpha kn\sigma J\mathbf{V}_n - \alpha(n\sigma)^2\mathbf{W}_n + 2\gamma\mathbf{U}_0\mathbf{U}_0^T\mathbf{W}_n &= \mathbf{0}. \end{aligned}$$

When the amplitude $\|\mathbf{U}_0\| = 0$ it is easy to show that all eigenvalues of the above system (i.e. taking solutions of the form $e^{\lambda t}$ and computing λ) are purely imaginary. Considering all other parameters fixed, and increasing $\|\mathbf{U}_0\|$, we find that there is a critical amplitude where the $n = 1$ mode becomes unstable through a collision of eigenvalues of opposite signature.

To analyze this instability, take $n = 1$ and study the reduced four dimensional system

$$J\partial_t \mathbf{V}_1 + 2\alpha k\sigma J\mathbf{W}_1 - \alpha\sigma^2 \mathbf{V}_1 + 2\gamma \mathbf{U}_0 \mathbf{U}_0^T \mathbf{V}_1 = \mathbf{0}$$
$$J\partial_t \mathbf{W}_1 - 2\alpha k\sigma J\mathbf{V}_1 - \alpha\sigma^2 \mathbf{W}_1 + 2\gamma \mathbf{U}_0 \mathbf{U}_0^T \mathbf{W}_1 = \mathbf{0}.$$

To determine the spectrum, let $(\mathbf{V}_1, \mathbf{W}_1) = (\mathbf{Q}, \mathbf{P})e^{\lambda t}$. Then $(\lambda, \mathbf{Q}, \mathbf{P})$ satisfy

$$\begin{bmatrix} \lambda J - \alpha\sigma^2 I + 2\gamma \mathbf{U}_0 \mathbf{U}_0^T & 2\alpha k\sigma J \\ -2\alpha k\sigma J & \lambda J - \alpha\sigma^2 I + 2\gamma \mathbf{U}_0 \mathbf{U}_0^T \end{bmatrix} \begin{pmatrix} \mathbf{Q} \\ \mathbf{P} \end{pmatrix} = \begin{pmatrix} \mathbf{0} \\ \mathbf{0} \end{pmatrix}. \qquad (4.8)$$

Denote the determinant of the left hand side by $\Delta(\lambda, \sigma)$. Then a calculation shows that

$$\Delta(\lambda, \sigma) = \lambda^4 + 2(p^2 + 4k^2\alpha^2\sigma^2)\lambda^2 + (p^2 - 4k^2\alpha^2\sigma^2)^2,$$

where

$$p^2 = \alpha^2\sigma^4 - 2\alpha\gamma\|\mathbf{U}_0\|^2\sigma^2.$$

Suppose $p^2 > 0$, then all four roots are purely imaginary (see Figure 3a) and given by

$$\lambda = \mathrm{i}2\alpha k\sigma \pm \mathrm{i}p, \quad -\mathrm{i}2\alpha k\sigma \pm \mathrm{i}p.$$

These modes are purely imaginary as long as $p^2 > 0$ or

$$2\gamma\alpha\|\mathbf{U}_0\|^2 < \alpha^2\sigma^2.$$

Since $\alpha\gamma > 0$, the instability threshold is achieved when the amplitude reaches

$$\|\mathbf{U}_0\| = \frac{|\alpha\sigma|}{\sqrt{2\alpha\gamma}}. \qquad (4.9)$$

At this threshold, a collision of eigenvalues occurs at the points $\lambda = \pm 2\mathrm{i}k\alpha\sigma$ (see Figure 3b).

Signature of the colliding modes Purely imaginary eigenvalues of a Hamiltonian system have a signature associated with them, and this signature is related to the sign of the energy (cf. Cairns (1979), MacKay and Saffman (1986), Bridges (1997)). Collision of eigenvalues of opposite signature is a necessary condition for the collision resulting in instability.

It is straightforward to compute the signature of the modes in the NLS model. Suppose that the amplitude $\|\mathbf{U}_0\|$ of the basic state is smaller than the critical value (4.9) for instability. Then there are two pairs of purely imaginary eigenvalues, and they each have a signature. Let us concentrate on the eigenvalues on the positive imaginary axis

$$\lambda = \mathrm{i}\Omega_\pm \quad \text{with} \quad \Omega_\pm = 2k\sigma\alpha \pm p. \qquad (4.10)$$

Then

$$\mathrm{Sign}(\Omega_\pm) = \mathrm{i}\langle \overline{\mathbf{Q}}, J\mathbf{Q}\rangle + \mathrm{i}\langle \overline{\mathbf{P}}, J\mathbf{P}\rangle,$$

where the inner product is real in order to make the conjugation explicit, and $(\mathbf{Q}, \mathbf{P})$ satisfy (4.8). One can also show that this signature has the same sign as the energy perturbation restricted to this mode.

The eigenvalue problem for $\Omega_\pm$ is (4.8) with λ replaced by $i\Omega_\pm$. From the system (4.8) it follows that

$$2\alpha k\sigma \mathbf{P} = \left[-i\Omega_\pm I - \alpha\sigma^2 J + 2\gamma J \mathbf{U}_0 \mathbf{U}_0^T\right] \mathbf{Q}.$$

First compute the case where $\|\mathbf{U}_0\| = 0$, then

$$2\alpha k\sigma \mathbf{P} = \left[-i\Omega_\pm I - \alpha\sigma^2 J\right] \mathbf{Q},$$

and so

$$(2\alpha k\sigma)^2 \mathsf{Sign}(\Omega_\pm) = i\left(2\Omega_\pm^2 \mp 2\alpha\sigma^2(2\alpha k\sigma)\right)\langle\overline{\mathbf{Q}}, J\mathbf{Q}\rangle - 2\alpha\Omega_\pm\sigma^2\langle\overline{\mathbf{Q}}, \mathbf{Q}\rangle.$$

After some computation it is found that

$$\mathbf{Q}_\pm = \begin{pmatrix} 1 \\ \pm i \end{pmatrix}, \quad \text{associated with the eigenvalue} \quad \lambda = i\Omega_\pm,$$

and so $\langle\overline{\mathbf{Q}}, \mathbf{Q}\rangle = 2$ and $\langle\overline{\mathbf{Q}}, J\mathbf{Q}\rangle = \mp 2i$. Substitution into the expression for signature yields

$$\mathsf{Sign}(\Omega_\pm) = \pm 4.$$

Hence the two modes which ultimately collide have opposite signature when $\mathbf{U}_0 = \mathbf{0}$. Since p^2 decreases as the amplitude increases, the two modes will have opposite signature for all $\|\mathbf{U}_0\|$ between $\|\mathbf{U}_0\| = 0$ and the point of collision.

5 The effect of dissipation on the Benjamin–Feir instability

Having established that there is a collision between modes of opposite signature, we can appeal to the well-known result that dissipation *always destabilizes* negative energy modes (Cairns (1979); Craik (1988); MacKay (1991)) to conclude that dissipation enhances the BF instability. But one has to be a bit careful. For example, Segur et al. (2004) recently studied the perturbed NLS equation with weak damping of the form

$$i u_t + \alpha\, u_{xx} + \gamma\, |u|^2 u + i\delta u = 0, \quad u \in \mathbb{C}, \tag{5.1}$$

where $\delta \geq 0$.

The perturbed NLS equation takes the form

$$J\mathbf{U}_t + \alpha\mathbf{U}_{xx} + \gamma\|\mathbf{U}\|^2\mathbf{U} + \delta J\mathbf{U} = \mathbf{0}, \tag{5.2}$$

or

$$J\mathbf{U}_t = \nabla H(\mathbf{U}) - \delta J\mathbf{U}, \tag{5.3}$$

with H defined by (3.4).

Looking for travelling wave solutions of this perturbed equation leads to

$$\left[(-\omega - \alpha k^2 + \gamma\|\mathbf{U}_0\|^2)I + \delta J\right]\mathbf{U}_0 = \mathbf{0}\,.$$

The main theoretical result of Segur et al. (2004) is that *a uniform train of plane waves of finite amplitude is stable for any physical system modeled by (5.1). Dissipation, no matter how small, stabilizes the Benjamin–Feir instability.*

The cornerstone of their argument is the δ−perturbed conservation law for $|u(x,t)|^2$,

$$\frac{\partial}{\partial t}|u|^2 = \mathrm{i}\alpha\frac{\partial}{\partial x}(\overline{u}u_x - u\overline{u}_x) - 2\delta|u|^2\,.$$

As above suppose $|u|^2$ can be integrated over an x−interval in such a way that $(\overline{u}u_x - u\overline{u}_x)$ vanishes at the boundary of the interval. Then

$$\frac{d}{dt}\int|u|^2\,\mathrm{d}x = -2\delta\int|u|^2\,\mathrm{d}x\,.$$

Clearly all solutions are damped no matter how small δ is. This argument is however a global argument. By integrating over space, all x−distribution and modal distribution of $|u|^2$ are smoothed out by the averaging (integration) process. From this global view it is impossible to determine the interplay between the various modes. It is shown below that in fact there is an interplay between the modes. Incidentally, it can be shown easily by using the δ−perturbed conservation law for the energy E (3.5) that the energy can grow with time (at best it is bounded from below).

A model problem   Consider the following prototype for a linear ordinary differential equation on $\mathbb{R}^4$ with a collision of eigenvalues of opposite signature, and Rayleigh damping,

$$\mathbf{Q}_{tt} + 2bJ\mathbf{Q}_t + (a - b^2)\mathbf{Q} + 2\delta\mathbf{Q}_t = \mathbf{0}\,, \tag{5.4}$$

where $\delta \geq 0$ represents the Rayleigh damping coefficient, a and b are real parameters and b is the "gyroscopic coefficient". Let

$$\mathbf{Q}(t) = \widehat{\mathbf{Q}}e^{\lambda t}\,,$$

then substitution into (5.4) leads to the characteristic polynomial

$$\Delta(\lambda) = (\lambda^2 + 2\delta\lambda - b^2 + a)^2 + 4b^2\lambda^2 = 0\,,$$

which can be factored into

$$[(\lambda + \mathrm{i}b)^2 + a + 2\delta\lambda] \times [(\lambda - \mathrm{i}b)^2 + a + 2\delta\lambda] = 0\,.$$

When $\delta = 0$ there are four roots $\lambda = \pm\mathrm{i}(b \pm a^{1/2})$. Suppose a is small and positive (just before the collision) and look at the effect of dissipation on the two modes

$$\lambda_0 = \mathrm{i}b \pm \mathrm{i}a^{1/2}\,.$$

Let

$$\lambda(\delta) - \lambda_0 + \delta\lambda_1 + 0(\delta^2).$$

Then substitution into the characteristic polynomial leads to

$$\lambda_1 = \mp\frac{1}{a^{1/2}}(b \pm a^{1/2}),$$

and so

$$\lambda(\delta) = ib \pm ia^{1/2} \mp \frac{\delta}{a^{1/2}}(b \pm a^{1/2}) + \mathcal{O}(\delta^2).$$

In other words, the negative energy mode $\lambda = i(b - a^{1/2})$ has positive real part when dissipatively perturbed, and the positive energy mode $\lambda = i(b + a^{1/2})$ has negative real part under perturbation. Consequently, when small dissipation is added the mode with negative energy will become unstable before the collision, and after the collision it will have a larger growth rate. This result is generic, and depends only on the property that the system has an instability arising through a collision of eigenvalues of opposite signature.

Effect of dissipation on the NLS model of BF instability Consider the $n = 1$ model reduced system for the BF instability with the type of damping introduced in (5.1) :

$$J\partial_t\mathbf{V}_1 + 2\alpha k\sigma J\mathbf{W}_1 - \alpha\sigma^2\mathbf{V}_1 + 2\gamma\mathbf{U}_0\mathbf{U}_0^T\mathbf{V}_1 + \delta J\mathbf{V}_1 = \mathbf{0}$$
$$J\partial_t\mathbf{W}_1 - 2\alpha k\sigma J\mathbf{V}_1 - \alpha\sigma^2\mathbf{W}_1 + 2\gamma\mathbf{U}_0\mathbf{U}_0^T\mathbf{W}_1 + \delta J\mathbf{W}_1 = \mathbf{0}.$$

This form of damping is very special. It can be factored out by the transformation

$$(\mathbf{V}_1, \mathbf{W}_1) = (\widetilde{\mathbf{V}_1}, \widetilde{\mathbf{W}_1})e^{-\delta t}.$$

Let's modify this damping slightly to break the symmetry, and to make it closer in spirit to Rayleigh damping, which is a much more widely used form of damping in mechanics.

Also without loss of generality, set $\|\mathbf{U}_0\| = 0$ since the negative energy and positive energy modes exist already then, and the analysis is exactly the same (but more complicated) for the case $\|\mathbf{U}_0\| > 0$,

$$J\partial_t\mathbf{V}_1 + 2\alpha k\sigma J\mathbf{W}_1 - \alpha\sigma^2\mathbf{V}_1 + \delta J\mathbf{V}_1 = \mathbf{0}$$
$$J\partial_t\mathbf{W}_1 - 2\alpha k\sigma J\mathbf{V}_1 - \alpha\sigma^2\mathbf{W}_1 + \delta J\mathbf{W}_1 + \frac{\delta^2}{2\alpha k\sigma}J\mathbf{V}_1 - \delta\frac{\sigma}{k}\mathbf{V}_1 = \mathbf{0}.$$

While the two extra terms in the second equation may look unusual, when the two equations are combined into a single second order equation we find

$$\partial_{tt}\mathbf{V}_1 + 2\alpha\sigma^2 J\partial_t\mathbf{V}_1 + (4\alpha^2 k^2\sigma^2 - \alpha^2\sigma^4)\mathbf{V}_1 + 2\delta\partial_t\mathbf{V}_1 = \mathbf{0},$$

which is precisely of the form (5.4), and so we can conclude immediately that dissipation *enhances* the BF instability in the NLS model! More details can be found in Bridges and Dias (2004).

6 Numerical integration of the NLS equation

The experimental and theoretical investigation of Lake et al. (1977) showed that the evolution of a 2D nonlinear wave train on deep water, in the absence of dissipative effects, exhibits the Fermi–Pasta–Ulam (FPU) recurrence phenomenon. This phenomenon is characterized by a series of modulation–demodulation cycles in which initially uniform wave trains become modulated and then demodulated until they are again uniform. As explained in Section 4, modulation is caused by the growth of the two dominant sidebands of the Benjamin–Feir instability at the expense of the carrier. During the demodulation the energy returns to the components of the original wave train (carrier, sidebands, harmonics). The FPU recurrence process is well described by the focusing 1D NLS equation. Lake et al. (1977) successfully solved the NLS equation numerically and found that the instability does not grow unboundedly as expected in linear theory, but that the conservation laws satisfied by the NLS equation tend to inhibit the growth. The result is that the solution returns to the initial condition periodically.

In the numerical study of the NLS equation there are two types of solutions which attract much interest : solitons, in which the solution and its spatial derivatives vanish at infinity (see for example Zakharov and Shabat (1971) for exact soliton solutions), and solutions that describe modulational instability and recurrence. Periodic boundary conditions are then appropriate. We focus on these solutions. Therefore we restrict ourselves to L-periodic solutions defined by

$$u(x + L, t) = u(x, t), \quad -\infty < x < \infty, \ t > 0. \tag{6.1}$$

Without loss of generality, we consider equation (3.1) with $\alpha = 1$. We have seen in Section 4 that an instability may occur if $\gamma > 0$. If

$$0 < n^2 \sigma^2 < 2\gamma \|\mathbf{U}_0\|^2,$$

two of the eigenvalues have positive real part and lead to exponential growth.

6.1   The split-step method

For this subsection, we follow closely the paper by Weideman and Herbst (1986). The solution of (3.1) may be advanced from one time level to the next by means of the formula

$$u(x, t + \tau) \approx \exp i\tau(\mathcal{L} + \mathcal{N}(u)) \cdot u(x, t), \tag{6.2}$$

where τ denotes the time step. In general (6.2) is first order accurate, but it is exact if $|u|^2$ is time-independent.

The time-splitting procedure consists of replacing the right-hand side of (6.2) by

$$\exp i\tau(\mathcal{L} + \mathcal{N}(u)) \cdot u(x, t) \approx \exp i\tau\mathcal{L} \cdot \exp i\tau\mathcal{N}(u) \cdot u(x, t).$$

This expression is exact whenever $\mathcal{L}$ and $\mathcal{N}$ commute. Otherwise the splitting is first order accurate. What the splitting does is to solve successively the equations

$$u_t = i\mathcal{N}(u)u, \quad u_t = i\mathcal{L}u,$$

where the solution of the former equation is used as initial condition for the latter.

From now on, U denotes the approximation to $u(x,t)$. Introducing the quantity

$$V^m = \exp i\tau \mathcal{N}(U^m) \cdot U^m \,, \tag{6.3}$$

where U^m denotes the approximation at the time $m\tau$, the split-step scheme can be written as

$$U^{m+1} = \exp i\tau \mathcal{L} \cdot V^m \,. \tag{6.4}$$

For the discretization of the space variable, one can use finite differences or a Fourier method. We give details for the Fourier method. The first step is to replace $V^m(x)$ and $U^{m+1}(x)$ in (6.4) by their Fourier series. The Fourier-series of a L-periodic function $w(x)$ is given by

$$w(x) = \sum_{n=-\infty}^{\infty} \hat{w}_n \exp(in\sigma x) \,,$$

with $\sigma = 2\pi/L$. The Fourier coefficients $\hat{w}_n$ are given by

$$\hat{w}_n = \frac{1}{L} \int_{-L/2}^{L/2} w(x) \exp(-in\sigma x) \mathrm{d}x \,.$$

This yields

$$\hat{u}_n^{m+1} = \exp(-in^2\sigma^2\tau)\hat{v}_n^m \,, \quad n \in \mathbb{Z} \,, \tag{6.5}$$

where the $\hat{v}_n^m$ and $\hat{u}_n^{m+1}$ are the Fourier coefficients of the continuous functions $V^m(x)$ and $U^{m+1}(x)$, respectively. The discretization of equation (6.5) gives, after replacing $\hat{v}_n^m$ by $\hat{V}_n^m$ and $\hat{u}_n^{m+1}$ by $\hat{U}_n^{m+1}$,

$$\hat{V}_n^m = \frac{1}{N} \sum_j V_j^m \exp(-in\sigma x_j) \,, \quad n = -N/2, \cdots, N/2 - 1$$

$$\hat{U}_n^{m+1} = \exp(-in^2\sigma^2\tau)\hat{V}_n^m \,, \quad n = -N/2, \cdots, N/2 - 1 \,.$$

The intermediate solution V_j^m is given by

$$V_j^m = \exp(i\tau\gamma|U_j^m|^2) \cdot U_j^m \,.$$

Finally the approximation at the next time level is calculated from the inverse transform

$$U_j^{m+1} = \sum_n \hat{U}_n^{m+1} \exp(in\sigma x_j) \,, \quad j = -N/2, \cdots, N/2 - 1 \,.$$

In the numerical computations, Fast Fourier Transforms are used.

Weideman and Herbst (1986) showed that a necessary condition for the solution to be stable again high frequency perturbation in the case $\gamma > 0$ is

$$\tau < \frac{L^2}{\pi N^2} \,.$$

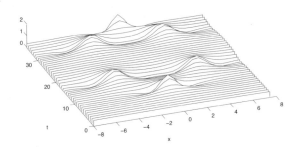

Figure 4. Illustration of the recurrence phenomenon. Split-step Fourier solution of the NLS equation with initial condition $u(x, 0) = 0.5[1 + 0.1(1 - |x|/8)]$, $-8 \leq x \leq 8$. The parameters are $\gamma = 2$, $L = 16$, $\tau = 0.09$. The number of points is $N = 30$.

In the case $\gamma < 0$, a necessary condition is

$$\frac{\tau}{2\theta + \pi} < \frac{L^2}{\pi^2 N^2} \,,$$

where

$$\theta = \sin^{-1} \left(\frac{\gamma \|\mathbf{U}_0\|^2 \tau}{[1 + \gamma^2 \|\mathbf{U}_0\|^4 \tau^2]^{1/2}} \right) .$$

These two conditions prevent any unwanted high frequency instabilities in the numerical solution.

6.2   Numerical results

We show some numerical results, based on the x-independent solution of the NLS equation

$$u(x, t) = \frac{1}{2} \exp \left(i \frac{\gamma t}{4} \right) ,$$

with the initial condition $u(x, 0) = \frac{1}{2}$. Perturbations of the initial condition will take the following form :

$$u(x, 0) = \frac{1}{2} \left[1 + 0.1(1 - 2|x|/L) \right] .$$

Energy is introduced into all modes.

In Figures 4 and 5, the only difference is the time step τ. The critical value for τ is $\tau^* = 0.0905$. In Figure 4, the numerical stability condition is satisfied ($\tau < \tau^*$). The only unstable modes are the two modes of lowest frequency. Most of the energy is transferred between the unstable modes with intermittent returns to the initial condition. In Figure 5 ($\tau > \tau^*$), there are in addition to the two analytically unstable modes corresponding to $|n| = 1$ and $|n| = 2$ a high frequency numerical instability corresponding to $|n| = 15$. After the return to the initial condition, the high frequency component suddenly becomes unstable, resulting in the highly oscillatory appearance of the solution.

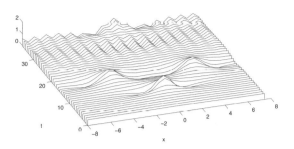

Figure 5. Same as Figure 4 with $\tau = 0.091$.

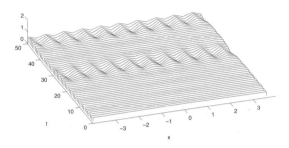

Figure 6. Same as Figure 4 with initial condition $u(x,0) = 0.5[1 + 0.1(1 - |x|/3.99)]$, $-3.99 \le x \le 3.99$. The parameters are $\gamma = -2$, $L = 7.98$, $\tau = 0.05$. The number of points is $N = 20$.

In Figure 6, $\gamma < 0$. If the numerical stability condition is satisfied, which is the case with $L = 8$, nothing happens as expected. If we take $L = 7.98$, numerical instability occurs. Figure 6 shows that the mode of highest frequency ($|n| = 10$) grows. But the solution returns to the initial state.

Some other numerical results are shown in Figures 7–9. The initial condition in Figure 7 leads to a homoclinic orbit. The initial conditions in Figures 8 and 9 are on opposite sides of the homoclinic orbit. See Ablowitz and Herbst (1990) for an explicit expression of the homoclinic solution and Osborne et al. (2000) for similar numerical results.

7 Bifurcation of waves when the phase and group velocities are nearly equal

As shown in Section 2, the dispersion relation for 2D capillary–gravity waves on the surface of a deep layer of water is given by

$$c^2 = g\frac{1}{k} + \frac{\sigma}{\rho}k. \tag{7.1}$$

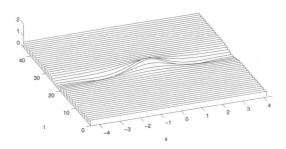

Figure 7. The initial condition is $u(x,0) = \frac{1}{2} + 10^{-5}(1 + i)\cos(2\pi x/L)$, with $L = 2\sqrt{2}\pi$. It leads to a homoclinic orbit. The parameters are $\gamma = 2$, $\tau = 0.05$. The number of points is $N = 30$.

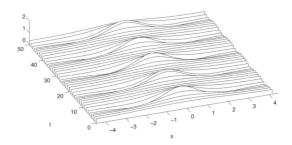

Figure 8. The initial condition is $u(x,0) = \frac{1}{2} + 0.1\cos(\sigma x)$, with $L = 2\sqrt{2}\pi$ and $\sigma = 2\pi/L$. The term $\cos(\sigma x)$ appears periodically. The other parameters are the same as in Figure 7.

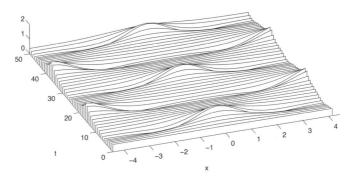

Figure 9. The initial condition is $u(x,0) = \frac{1}{2} + 0.1\,i\cos(\sigma x)$, with $L = 2\sqrt{2}\pi$ and $\sigma = 2\pi/L$. Interchanges between $\cos(\sigma x)$ and $\cos\sigma(x + \frac{1}{2}L)$ take place.

A trivial property of this dispersion relation is that it exhibits a minimum $c_{\min}$, but surprisingly it was only quite recently that some of its consequences were discovered. In the classic textbooks such as Lamb (1932), pp. 462–468; Lighthill (1978), pp. 260–269; Whitham (1974), pp. 407–408, pp. 446–454; Milne-Thomson (1968), pp. 447–449; Stoker (1958), the presence of this minimum is of course mentioned. It represents a real difficulty for linearized versions of the water-wave problem. For example, consider the fishing-rod problem, in which a uniform current is perturbed by an obstacle. Rayleigh (1883) investigated this problem. He assumed a distribution of pressure of small magnitude and linearized the equations around a uniform stream with constant velocity c. He solved the resulting linear equations in closed form. For $c = c_1 (> c_{\min})$, the solutions are characterized by trains of waves in the far field of wavenumbers k_2 and $k_1 < k_2$. The waves corresponding to k_1 and k_2 appear behind and ahead of the obstacle, respectively. The asymptotic wave trains are given by

$$\eta \quad \sim \quad -\frac{2\rho P}{(k_2 - k_1)\sigma} \sin(k_1 x), \quad x > 0,$$

$$\eta \quad \sim \quad -\frac{2\rho P}{(k_2 - k_1)\sigma} \sin(k_2 x), \quad x < 0,$$

where P is the integral of pressure. For $c < c_{\min}$, Rayleigh's solutions do not predict waves in the far field, and the flow approaches a uniform stream with constant velocity c at infinity. This is consistent with the fact that equation (7.1) does not have real roots for k when $c < c_{\min}$. Rayleigh's solution is accurate for $c \neq c_{\min}$ in the limit as the magnitude of the pressure distribution approaches zero. However it is not uniform as $c \to c_{\min}$. It is clear that the linearized theory fails as one approaches $c_{\min}$: the two wavenumbers k_1 and k_2 merge, the denominators approach zero, and the displacement of the free surface becomes unbounded. Therefore there was a clear need for a better understanding of the limiting process, but for several decades this problem was left untouched. In the late 80's and early 90's, several researchers worked on the nonlinear version of this problem independently. Longuet-Higgins (1989) indirectly touched upon this problem with numerical computations. Iooss and Kirchgässner (1990) tackled the problem mathematically. Vanden-Broeck and Dias (1992) made the link between the numerical computations of Longuet-Higgins (1989) and the mathematical analysis of Iooss and Kirchgässner (1990). One can say that the mathematical results shed some light on the difficulty : there is a difference between a temporal approach and a spatial approach. Roughly speaking, in temporal bifurcation theory the wavenumber k is treated as a given real parameter while in spatial bifurcation theory the wavespeed c is treated as a given real parameter. Again the best way to understand what happens is to consider a model equation. The appropriate model equation is the NLS equation for the amplitude A of a modulated wave train :

$$2\mathrm{i}\frac{\partial A}{\partial t_2} + p\frac{\partial^2 A}{\partial X^2} + q\frac{\partial^2 A}{\partial y_1^2} + \gamma A|A|^2 = 0. \tag{7.2}$$

Akylas (1993) and Longuet-Higgins (1993) showed that, for values of c less than $c_{\min}$, equation (7.2) admits particular envelope-soliton solutions, such that the wave crests are stationary in the reference frame of the wave envelope. These solitary waves, which bifurcate from linear periodic waves at the minimum value of the phase speed, have decaying oscillatory tails and are sometimes called "bright" solitary waves. More generally, one can look for stationary solutions of equation (7.2). Using $\sigma/\rho c^2$ as unit length and $\sigma/\rho c^3$ as unit time, allowing for interfacial waves (i.e. waves propagating at the interface between a heavy fluid of density ρ_1 and a lighter fluid of density ρ_2), considering waves without y_1-variations and evaluating the coefficients p and γ at $c = c_{\min}$, one can show that these stationary solutions satisfy the equation

$$-\frac{2r}{1+r}\mu A + A_{XX} + \frac{16r^2 - 5}{2(1+r)^2} A|A|^2 = 0 \,, \tag{7.3}$$

where the bifurcation parameter μ is defined by $\mu = \alpha - \alpha_{\min}$, with $\alpha = g\sigma/\rho c^4$ and $\alpha_{\min} = g\sigma/\rho c_{\min}^4 = 1/2r(1+r)$. The parameter r is the density ratio $(\rho_1 - \rho_2)/(\rho_1 + \rho_2)$. Note that the coefficient of the cubic term vanishes when $r = r_0 = \sqrt{5}/4$. The corresponding profile for the modulated wave is given by

$$\eta(X) = (1+r)\left[A(X)\exp\left(iX/(1+r)\right) + \text{c.c.}\right].$$

Introduce the scaling $|\mu|^{\frac{1}{2}}\tilde{A} = A$, $\tilde{X} = |\mu|^{\frac{1}{2}}\left(2r/(1+r)\right)^{\frac{1}{2}} X$, and the coefficient

$$\tilde{\gamma} = \frac{16r^2 - 5}{4r(1+r)} \,.$$

The resulting equation is

$$\mathsf{Sign}(\mu)\tilde{A} - \tilde{A}_{\tilde{X}\tilde{X}} - \tilde{\gamma}\tilde{A}|\tilde{A}|^2 = 0 \,. \tag{7.4}$$

Writing $\tilde{A} = s(\tilde{X})\,e^{i\theta(\tilde{X})}$ leads to

$$s_{\tilde{X}\tilde{X}} - \mathsf{Sign}(\mu)s + \tilde{\gamma}s^3 - s(\theta_{\tilde{X}})^2 = 0 \,, \tag{7.5}$$
$$2\theta_{\tilde{X}}s_{\tilde{X}} + s\theta_{\tilde{X}\tilde{X}} = 0 \,. \tag{7.6}$$

The system has two first integrals I_1 and I_2, defined as follows :

$$u\theta_{\tilde{X}} = I_1 \,, \tag{7.7}$$
$$\tfrac{1}{4}(u_{\tilde{X}})^2 = \mathsf{Sign}(\mu)u^2 - \tfrac{1}{2}\tilde{\gamma}u^3 - I_1^2 + I_2 u \,, \tag{7.8}$$

where $u \equiv s^2$. These two integrals are related to the energy flux and flow force, respectively, as shown by Bridges et al. (1995).

For a full description of all the bounded solutions of equation (7.4), one can refer to Iooss and Pérouème (1993), Dias and Iooss (1993) or Dias and Iooss (1996). There are four cases to consider :

- $r < r_0$, $c > c_{\min}$: there are periodic solutions, quasiperiodic solutions and solitary waves, homoclinic to the same periodic wave with a phase shift at $+\infty$ and $-\infty$ (these homoclinic solutions are sometimes called "dark" solitary waves if the amplitude vanishes at the origin and "grey" solitary waves if it does not). A dark solitary wave is shown in Figure 10.

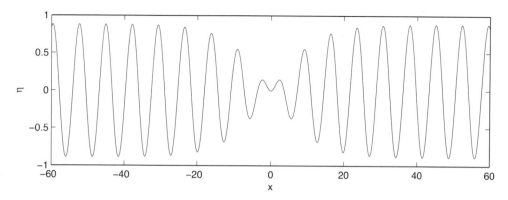

Figure 10. Profile of a dark solitary wave, given by (7.9), with $r = 0.15$ and $\mu = 0.05$.

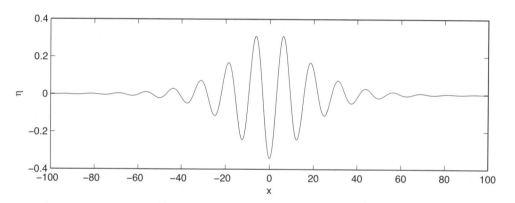

Figure 11. Profile of a bright solitary wave, given by (7.10), with $\mu = 0.005$.

- $r < r_0$, $c < c_{\min}$: there are no bounded solutions.

- $r > r_0$, $c > c_{\min}$: there are periodic solutions and quasiperiodic solutions.

- $r > r_0$, $c < c_{\min}$: there are periodic solutions (of finite amplitude only), quasiperiodic solutions and solitary waves, homoclinic to the rest state (such a solitary wave is shown in Figure 11).

When $r < r_0$ (defocussing case), $c > c_{\min}$ ($\mu < 0$), there is a one-parameter family of grey solitary waves (Dias and Iooss (1996), Laget and Dias (1997)). The "darkest" one, which is such that the amplitude vanishes at the origin, has an envelope given by

$$s = |\tilde{\gamma}|^{-\frac{1}{2}} \tanh\left(\frac{|\tilde{X}|}{\sqrt{2}}\right).$$

The elevation of the dark solitary wave is given by

$$\eta(X) = \pm 4 \sqrt{\frac{r(1+r)^3}{5 - 16r^2}} \tanh\left(\sqrt{\frac{r}{1+r}} \, |\mu|^{\frac{1}{2}} X\right) \sin\left(\frac{1}{1+r} X\right). \tag{7.9}$$

When $r > r_0$ (focussing case), $c < c_{\min}$ ($\mu > 0$), there are bright solitary waves, the envelope of which is given by

$$\tilde{A} = \pm \frac{\sqrt{2}}{\sqrt{\tilde{\gamma}} \cosh \tilde{X}}.$$

This case includes water waves. The elevation of the solitary wave for water waves is given by

$$\eta(X) = \pm \frac{16\sqrt{\mu}}{\sqrt{11}} \frac{\cos(X/2)}{\cosh \sqrt{\mu}X}. \tag{7.10}$$

The results obtained on the NLS equation also apply to the full water-wave and interfacial wave problems. In particular envelope-soliton solutions have been studied in detail by Longuet-Higgins (1989), Vanden-Broeck and Dias (1992), Dias et al. (1996) and Dias and Iooss (1993).

Supporting the asymptotic and numerical studies cited above, Iooss and Kirchgässner (1990) provided a rigorous proof, based on center-manifold reduction, for the existence of small-amplitude symmetric solitary waves near the minimum phase speed in water of finite depth. The proof could not be extended to the infinite-depth case, however. Later, Iooss and Kirrmann (1996) handled this difficulty by following a different reduction procedure which also brought out the fact that the solitary-wave tails behave differently in water of infinite depth, their decay being slower than exponential, although the precise decay rate could not be determined. By assuming the presence of an algebraic decay, Sun (1997) was able to show that the profiles of interfacial solitary waves in deep fluids must decay like $1/x^2$ at the tails. We also remark that earlier Longuet-Higgins (1989) had inferred such a decay on physical grounds for deep-water solitary waves. Akylas, Dias and Grimshaw (1998) showed that the profile of these gravity–capillary solitary waves actually decays algebraically (like $1/x^2$) at infinity, owing to the induced mean flow that is not accounted for in the NLS equation.

The NLS equation also admits asymmetric solitary waves, obtained by shifting the carrier oscillations relative to the envelope of a symmetric solitary wave. Yang and Akylas (1997) examined the fifth-order Korteweg–de Vries equation, a model equation for gravity–capillary waves on water of finite depth, and showed by using techniques of exponential asymptotics beyond all orders that asymmetric solitary waves are not possible. On the other hand, an infinity of symmetric and asymmetric solitary waves, in the form of two or more NLS solitary wavepackets, exist at finite amplitude.

The implications of the study of the 1:1 resonance in the context of water waves have gone far beyond the field of surface waves. Applications have been given to all sorts of problems in physics, mechanics, thermodynamics and optics since these studies on water waves. See Dias and Iooss (2003) for a mathematical review.

8 The 2D "hyperbolic" NLS equation

As shown in Section 2, the 2D NLS equation that arises in the description of surface gravity waves on deep water is

$$i\partial_t\psi + \partial_{xx}\psi - \partial_{yy}\psi + |\psi|^2\psi = 0 \,. \tag{8.1}$$

This is the two-dimensional "hyperbolic" Schrödinger equation. Sulem and Sulem (1999) write that numerical integrations of (8.1) in spatially periodic domains show no tendency to collapse, and FPU recurrence is observed. However, in contrast with the focusing one-dimensional problem, this recurrence can only be approximate, since a small fraction of the energy is pumped to higher frequencies during each cycle because the region of unstable wave vectors is unbounded. In contrast, for the two-dimensional elliptic NLS equation, numerical evidence of finite-time blowup has been reported.

Osborne et al. (2000) write that the very existence of "coherent structures" in the 2D hyperbolic NLS equation (or "unstable modes" in the sense of the 1D NLS equation) has been left in doubt. They performed more numerical computations and concluded that unstable modes do indeed exist in the 2D hyperbolic NLS equation. They can take the form of large amplitude "rogue" waves. Onorato et al. (2001) extended these results to random initial conditions and concluded that freak waves are more likely to occur for large values of the Phillips parameter and the enhancement coefficient in the Joint North Sea Wave Project (JONSWAP) power spectrum.

9 Forced NLS equation

The NLS equation can still be used as a model when one considers water waves forced by an obstacle or by a pressure disturbance on the free surface. Let us for example consider capillary–gravity waves forced by a pressure distribution acting locally on the free surface and moving at a velocity slightly below the velocity corresponding to the minimum of the dispersion curve. In this regime, the linear analysis of Rayleigh (1883) is no longer valid.

The water wave problem has several formulations. One based on the velocity potential was presented in Section 2. One can also use a formulation based on the conservation of mass and the irrotationality condition. In 2D, the resulting equations are

$$u_x + w_z = 0, \quad u_z - w_x = 0 \,. \tag{9.1}$$

In addition to the problem considered in Section 2, there is now a pressure distribution moving along the free surface at the speed c. Introduce nondimensional variables by taking c as unit velocity and $\sigma/\rho c^2$ as unit length. The boundary conditions in a frame of reference moving with the pressure distribution are given on the bottom by

$$w = 0, \quad \text{for } z = -1/b$$

and on the free-surface by

$$u\eta_x - w = 0, \quad \frac{1}{2}(u^2 + w^2 - 1) + b\lambda\eta - \frac{\eta_{xx}}{(1+\eta_x^2)^{3/2}} + \varepsilon p_0 = 0, \quad \text{for } z = \eta(x). \tag{9.2}$$

Here $\lambda = gh/c^2$ and $b = \sigma/\rho hc^2$ are dimensionless numbers. The dimensionless pressure is denoted by $\varepsilon p_0(x)$. The function p_0 is assumed to be of compact support and such that $\langle p_0 \rangle = \int_{-\infty}^{\infty} p_0(x)dx \neq 0$. The dispersion relation for capillary–gravity waves is given by

$$(b\lambda + k^2)\tanh(k/b) - k = 0, \tag{9.3}$$

where k is the dimensionless wavenumber. For any $k \in [0, 1/2]$, there exists a unique pair $\lambda(k), b(k)$ such that equation (9.3) admits a double root in k. The corresponding speed is $c_{\min}$.

The analysis is based on reformulating equations (9.1)-(9.2) as a dynamical system in x, on reducing the problem to its center manifold as in Kirchgässner (1988) and on putting the reduced system in normal form in the presence of a reversible 1:1 resonance as in Iooss and Pérouème (1993). Introducing $v_1 = (u^2 + w^2 - 1)/2$, $v_2 = w/u$, v_0 the trace of v_2 on the free surface, $\varphi_0^\pm, \varphi_1^\pm$ the eigenvectors and generalized eigenvectors corresponding to the double imaginary eigenvectors $\pm ik$, and the parameter μ proportional to $c_{\min} - c$, one can write

$$(v_0, v_1, v_2) = A(x)\varphi_0^+ + B(x)\varphi_1^+ + \bar{A}(x)\varphi_0^- + \bar{B}(x)\varphi_1^- + \Phi(\mu, \varepsilon; x, A, B, \bar{A}, \bar{B}), \tag{9.4}$$

with A and B given by the normal form

$$A_x = ikA + B + iAP\left(\mu; |A|^2, \tfrac{1}{2}i(A\bar{B} - \bar{A}B)\right) - i\mathcal{P}_1\mathcal{P}_0\varepsilon p_0 + \cdots \tag{9.5}$$

$$B_x = ikB + iBP\left(\mu; |A|^2, \tfrac{1}{2}i(A\bar{B} - \bar{A}B)\right) + AQ\left(\mu; |A|^2, \tfrac{1}{2}i(A\bar{B} - \bar{A}B)\right)$$
$$+ \mathcal{P}_0\varepsilon p_0 + \cdots, \tag{9.6}$$

where P and Q are polynomials defined as

$$P(\mu; U, V) = p_1\mu + p_2U + p_3V + \mathcal{O}(|\mu| + |U| + |V|)^2, \tag{9.7}$$

$$Q(\mu; U, V) = q_1\mu - q_2U + q_3V + \mathcal{O}(|\mu| + |U| + |V|)^2. \tag{9.8}$$

The positive constants $\mathcal{P}_0$ and $\mathcal{P}_1$ have been computed by Dias and Iooss (1993). After rescaling, the system (9.5) can be replaced by

$$\tilde{A}_{\tilde{x}} = \tilde{B} + \mathcal{O}\left(|\mu|^{1/2}\right) \tag{9.9}$$

$$\tilde{B}_{\tilde{x}} = \tilde{A}(q_1\text{Sign}(\mu) - q_2|\tilde{A}|^2) + \tilde{\varepsilon}e^{-ikx}\frac{P_0}{\langle P_0 \rangle} + \mathcal{O}\left(|\mu|^{1/2}\right). \tag{9.10}$$

The sign of $\tilde{\varepsilon}$ depends on the sign of the pressure distribution. One can replace $P_0/\langle P_0 \rangle$ by δ_0, the Dirac delta function, since $P_0/\langle P_0 \rangle$ converges towards 0 for all $\tilde{x} \neq 0$ and the average $\langle P_0/\langle P_0 \rangle \rangle$ is equal to 1. At leading order, one gets the forced nonlinear Schrödinger equation

$$\tilde{A}_{\tilde{x}\tilde{x}} = q_1\text{Sign}(\mu)\tilde{A} - q_2\tilde{A}|\tilde{A}|^2 + \tilde{\varepsilon}\delta_0. \tag{9.11}$$

The mathematical problem of the persistence of the solutions of equation (9.11) when dealing with the full system (9.1)-(9.2) can be tackled as in Iooss and Pérouème (1993) or Iooss and Kirrmann (1996) for the study of persistence without forcing, and Mielke

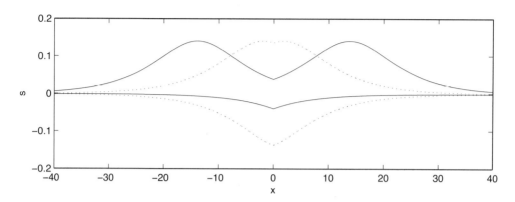

Figure 12. Forced gravity–capillary solitary waves in deep water. The four possibilities for the wave envelope $s(x)$ given by equation (9.12).

(1986) or Kirchgässner (1988) for the study of persistence with forcing. But there is numerical evidence by Vanden-Broeck and Dias (1992) that the solutions obtained here persist.

We look for continuous and bounded solutions of equation (9.11) on the whole real line, with $\tilde{A}_{\tilde{x}}(0+) - \tilde{A}_{\tilde{x}}(0-) = \tilde{\varepsilon}$, which satisfy (9.11) with $\tilde{\varepsilon} = 0$ for $\tilde{x} \neq 0$. The analysis is restricted to homoclinic solutions. Let $\mu_0 = (q_2/2q_1^2)^{1/2} \varepsilon \mathcal{P}_0 \langle p_0 \rangle$. When $0 < \mu < \mu_0$, there are no homoclinic solutions. When $\mu = \mu_0$, there are two homoclinic solutions. When $\mu > \mu_0$, there are four homoclinic solutions. Writing $A(x) = s(x)e^{i(kx+\theta(x))}$, we show in Figure 12 the four possibilities for the envelope $s(x)$ in deep water. The corresponding wave profiles $\eta(x) = 4\mathrm{Re}(s(x)e^{ikx})$ are shown in Figure 13. The expression of $s(x)$ is given by

$$s(x) = \pm(2q_1\mu/q_2)^{1/2}/\cosh(\sqrt{q_1\mu}|x| \pm \alpha_{1/2}), \tag{9.12}$$

where $\alpha_{1/2} > 0$ are the roots of $\sinh\alpha/\cosh^2\alpha = \mu_0/2\mu$. Figure 14 shows the amplitude $|\eta(0)|$ as a function of μ for : (i) the solution of the linearized problem with $\varepsilon \neq 0$, (ii) the analytical solution with $\varepsilon = 0$, (iii) the present solutions with $\varepsilon \neq 0$. More details can be found in Pǎrǎu and Dias (2000).

Forced nonlinear Schrödinger equations have been obtained elsewhere (see for example Akylas (1984) and Barnard et al. (1977)). Even if $q_2 < 0$, $q_1 > 0$, $\mu > 0$, one can obtain homoclinic solutions to zero, although it is impossible without forcing as shown for example by Pǎrǎu and Dias (2002). Our goal here was to deal with the divergence of the solutions of the linearized problem when the velocity c approaches $c_{\min}$, by studying the influence of the nonlinear terms.

Bibliography

M.J. Ablowitz and B.M. Herbst, On homoclinic structure and numerically induced chaos for the nonlinear Schrödinger equation. *SIAM Journal of Applied Mathematics*, 50:339–351, 1990.

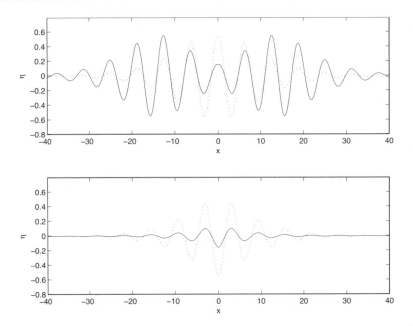

Figure 13. Forced gravity–capillary solitary waves in deep water. The corresponding profiles $\eta(x)$ with $s(x) > 0$ (top) and with $s(x) < 0$ (bottom).

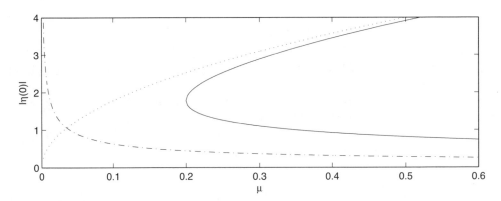

Figure 14. Amplitude at the origin as a function of μ. Comparison between linearized theory with pressure (dash-dotted line), nonlinear theory without pressure (dotted line) and nonlinear theory with pressure (solid line).

T. Akylas, On the excitation of nonlinear water waves by a moving pressure distribution oscillating at resonant frequency. *Phys. Fluids*, 27:2803–2807, 1984.

T. Akylas, Envelope solitons with stationary crests. *Phys. Fluids A*, 5:789–791, 1993.

T. Akylas, F. Dias, and R. Grimshaw, The effect of the induced mean flow on solitary waves in deep water. *J. Fluid Mech.*, 355:317–328, 1998.

J.S. Barnard, J.J. Mahony, and W.G. Pritchard, The excitation of surface waves near a cut-off frequency. *Phil. Trans. R. Soc. London A*, 286:87–123, 1977.

T.B. Benjamin and J.E. Feir, The disintegration of wavetrains in deep water. Part 1. *J. Fluid Mech.*, 27:417–430, 1967.

D.J. Benney and A.C. Newell, The propagation of nonlinear wave envelopes. *J. Math. and Phys.*, 46:133–139, 1967.

T.J. Bridges, A geometric formulation of the conservation of wave action and its implications for signature and the classification of instabilities. *Proc. Roy. Soc. London Ser. A*, 453:1365–1395, 1997.

T. Bridges, P. Christodoulides, and F. Dias, Spatial bifurcations of interfacial waves when the phase and group velocities are nearly equal. *J. Fluid Mech.*, 295:121–158, 1995.

T. J. Bridges and F. Dias, On the enhancement of the Benjamin–Feir instability due to dissipation. *preprint*, 2004.

T.J. Bridges and A. Mielke, A proof of the Benjamin–Feir instability. *Arch. Rat. Mech. Anal.*, 133:145–198, 1995.

R.A. Cairns, The role of negative energy waves in some instabilities of parallel flows. *J. Fluid Mech.*, 92:1–14, 1979.

T. Colin, F. Dias, and J.-M. Ghidaglia, On rotational effects in the modulations of weakly nonlinear water waves over finite depth. *Eur. J. Mech. B/Fluids*, 14:775–793, 1995.

W. Craig, C. Sulem, and P.L. Sulem, Nonlinear modulation of gravity waves: a rigorous approach. *Nonlinearity*, 5:497–522, 1992.

A.D.D. Craik, *Wave Interactions and Fluid Flows*. Cambridge University Press, 1988.

F. Dias and G. Iooss, Capillary–gravity solitary waves with damped oscillations. *Physica D*, 65:399–423, 1993.

F. Dias and G. Iooss, Capillary–gravity interfacial waves in deep water. *Europ. J. Mech. B*, 15:367–390, 1996.

F. Dias and G. Iooss, Water waves as a spatial dynamical system. In *Handbook of Mathematical Fluid Dynamics*, Vol. 2, Ed. S. Friedlander and D. Serre, Elsevier, pages 443–449, 2003.

F. Dias and C. Kharif, Nonlinear gravity and capillary–gravity waves. *Annu. Rev. of Fluid Mech.*, 31:301–346, 1999.

F. Dias, D. Menasce, and J.M. Vanden-Broeck, Numerical study of capillary–gravity solitary waves. *Europ. J. Mech. B/Fluids*, 15:17–36, 1996.

K.B. Dysthe, Note on a modification to the nonlinear Schrödinger equation for application to deep water waves. *Proc. R. Soc. Lond. A*, 369:105–114, 1979.

J.L. Hammack and D.M. Henderson, Resonant interactions among surface water waves. *Annu. Rev. Fluid Mech.*, 25:55–97, 1993.

S.J. Hogan, The fourth-order evolution equation for deep-water gravity–capillary waves. *Proc. R. Soc. Lond. A*, 402:359–372, 1985.

G. Iooss and K. Kirchgässner, Bifurcation d'ondes solitaires en présence d'une faible tension superficielle. *C. R. Acad. Sci. Paris, Série I*, 311:265–268, 1990.

G. Iooss and P. Kirrmann, Capillary gravity waves on the free surface of an inviscid fluid of infinite depth. Existence of solitary waves. *Arch. Rat. Mech. Anal.*, 136:1–19, 1996.

G. Iooss and M.-C. Pérouème, Perturbed homoclinic solutions in reversible 1:1 resonance fields. *Journal of Differential Equations*, 102:62–88, 1993.

K. Kirchgässner, Nonlinearly resonant surface waves and homoclinic bifurcation. *Adv. Applied Mech.*, 26:135–181, 1988.

V.P. Krasitskii, Canonical transformation in a theory of weakly nonlinear waves with a nondecay dispersion law. *Sov. Phys. JETP*, 71:921–927, 1990.

O. Laget and F. Dias, Numerical computation of capillary–gravity interfacial solitary waves. *J. Fluid Mech.*, 349:221–251, 1997.

B.M. Lake, H.C. Yuen, H. Rungaldier, and W.E. Ferguson, Nonlinear deep-water waves : theory and experiment. Part 2. Evolution of a continuous wave train. *J. Fluid Mech.*, 83:49–74, 1977.

H. Lamb, *Hydrodynamics*. Dover Publications, 1932.

M.J. Lighthill M. J., *Waves in fluids*. Cambridge University Press, 1978.

M. Longuet-Higgins, Capillary–gravity waves of solitary type on deep water. *J. Fluid Mech.*, 200:451–70, 1989.

M.S. Longuet-Higgins, Capillary–gravity waves of solitary type and envelope solitons on deep water. *J. Fluid Mech.*, 252:703–711, 1993.

R.S. MacKay, Movement of eigenvalues of Hamiltonian equilibria under non-Hamiltonian perturbation. *Phys. Lett. A*, 155:266–268, 1991.

R.S. MacKay and P. Saffman, Stability of water waves. *Proc. Roy. Soc. London Ser. A*, 406:115–125, 1986.

A. Mielke, Steady flows of inviscid fluids under localized perturbations. *J. Diff. Eq.*, 65:89–116, 1986.

L.M. Milne-Thomson, *Theoretical hydrodynamics*. Dover Publications, 1968.

M. Onorato, A.R. Osborne, M. Serio, and S. Bertone, Freak waves in random oceanic sea states. *Phys. Rev. Lett.*, 86:5831–5834, 2001.

A.R. Osborne, M. Onorato, and M. Serio, The nonlinear dynamics of rogue waves and holes in deep-water gravity wave trains. *Phys. Lett. A*, 275:386–393, 2000.

E. Pǎrǎu and F. Dias, Ondes solitaires forcées de capillarité–gravité. *C. R. Acad. Sci. Paris I*, 331:655–660, 2000.

E. Pǎrǎu and F. Dias, Nonlinear effects in the response of a floating ice plate to a moving load. *J. Fluid Mech.*, 460:281–305, 2002.

D.H. Peregrine, Water waves, nonlinear Schrödinger equations and their solutions. *J. Aust. Math. Soc. B*, 25:16–43, 1983.

O.M. Phillips, Wave interactions – the evolution of an idea. *J. Fluid Mech.*, 106:215–227, 1981.

O.M. Rayleigh Lord, *Proc. London Math. Soc.*, 15:69, 1883.

G. Schneider, Approximation of the Korteweg-de Vries equation by the nonlinear Schrödinger equation. *Journal of Differential Equations*, 147:333–354, 1998.

H. Segur, D. Henderson, J. Hammack, C.-M. Li, D. Pheiff, and K. Socha, Stabilizing the Benjamin–Feir instability. *preprint*, 2004.

M. Stiassnie, Note on the modified nonlinear Schrödinger equation for deep water waves. *Wave Motion*, 6:431–433, 1984.

J.J. Stoker, *Water waves, the mathematical theory with applications*. Wiley-Interscience, 1958.

C. Sulem and P.-L. Sulem, *The nonlinear Schrödinger equation*. Springer, 1999.

S.M. Sun, Some analytical properties of capillary–gravity waves in two-fluid flows of infinite depth. *Proc. R. Soc. London Ser. A*, 453:1153–1175, 1997.

J.-M. Vanden-Broeck and F. Dias, Gravity–capillary solitary waves in water of infinite depth and related free-surface flows. *J. Fluid Mech.*, 240:549–557, 1992.

J.A.C. Weideman and B.M. Herbst, Split-step methods for the solution of the nonlinear Schrödinger equation. *SIAM Journal of Numerical Analysis*, 23:485–507, 1986.

G.B. Whitham, *Linear and nonlinear waves*. Wiley-Interscience, 1974.

T.S. Yang and T.R. Akylas, On asymmetric gravity–capillary solitary waves. *J. Fluid Mech.*, 330:215–232, 1997.

V.E. Zakharov, Stability of periodic waves of finite amplitude on the surface of a deep fluid. *Zh. Prikl. Mekh. Tekh. Fiz.*, 9:86–94, 1968 (Transl. in *J. Appl. Mech. Tech. Phys.*, 9:190–194, 1968).

V.E. Zakharov and A.B. Shabat, Exact theory of two-dimensional self-modulating waves in non-linear media. *Zh. Eksp. Teor. Fiz.*, 61:118, 1971 (Transl. in *J. Exp. Theor. Phys.*, 34:62–69, 1972).

Wave Interactions

J. Vanneste

School of Mathematics, University of Edinburgh, Edinburgh UK

Abstract. Waves with different wavenumbers and frequencies interact when they propagate in a nonlinear medium. In the weakly nonlinear limit, the interactions involve only small sets of waves (triads or quartets), and they are governed by simple amplitude equations. These lectures present the derivation of these amplitude equations, examine their properties and solutions, and discuss some applications to waves in fluids, in particular to Rossby waves and surface gravity waves.

1 Introduction

When studying small-amplitude motion in fluids, the first approximation that is naturally made is to neglect the nonlinear terms in the equations of motion. Assuming that the resulting linear equations are time and space independent, solutions can be sought in the form of plane waves, with structure proportional to $\exp[i(\mathbf{k} \cdot \mathbf{x} - \omega t)]$, where ω is a constant frequency related to the constant wavevector $\mathbf{k}$ by the dispersion relation. By superposition, these plane waves with constant amplitude can be used to express the general solution of the linearized equations of motion.

Nonlinear effects, when they are weak enough, do not fundamentally alter the nature of the motion. A weakly nonlinear solution can still be usefully expressed as a superposition of plane waves, but the amplitudes of these waves do not remain constant: they are modulated by nonlinear interactions. Over long time scales, typically inversely proportional to the wave amplitudes, these modulations can lead to order-one changes in the amplitudes. Simple geometric constraints imply that interactions between waves are restricted to particular sets, such that the sum of the wavevectors vanishes; furthermore, the interactions are genuinely significant (in the sense that they lead to order-one amplitude changes) only if the waves involved are resonant, that is, if the sum of their frequencies vanishes. Assuming special initial conditions, it is interesting (and justified) to focus on the interactions within the smallest possible sets of resonantly interacting waves; typically these will be sets of three waves — wave triads — or for some waves (such as surface gravity waves) sets of four waves — wave quartets. In these lectures, we review the properties of these interactions.

Motivated by applications to inviscid fluids, we concentrate on waves propagating in conservative systems (Hamiltonian or Lagrangian systems). The structure that the associated conservation laws impose is sufficient to characterize the wave interactions completely; thus, the wave interactions are essentially identical for all conservative systems (although the dispersion relation is crucial in determining the possible resonant sets). Here, we use the examples of Rossby waves and internal gravity waves for triad interactions, and of surface gravity waves for quartet interactions to illustrate our relatively general formulation.

Given the many physical contexts to which they are relevant, it is not surprising that wave interactions have generated a vast literature. Since the 1960s, numerous results have been derived, usually for particular types of waves to start with (often in plasma physics) before being applied to other waves, in particular to waves in fluids. These lectures present a small selection of these results; we refer to the books of Craik (1985) and Ichimaru (1973) for further results and references.

2 Linear waves in conservative systems

Plane waves naturally emerge as basic solutions of linearized equations of motion. They provide the building blocks with which one can examine many processes such as dispersion and, in these lectures, weakly nonlinear interactions. We start by reviewing some essential properties of plane waves in conservative, i.e. energy- and momentum-conserving systems. We first consider the example of Rossby waves before providing a more general formulation.

2.1   Linear Rossby waves

Rossby waves are large-scale waves propagating in the atmosphere and the oceans. To describe these waves in the simplest manner, we model the atmosphere or the oceans as a two-dimensional fluid on the so-called β-plane which accounts for the North–South variation in the Coriolis force. The dynamics is governed by the vorticity equation

$$\zeta_t + \beta\psi_x + \psi_x\zeta_y - \psi_y\zeta_x = 0. \tag{2.1}$$

Here ψ is a streamfunction from which the velocity field $(u, v) = (-\psi_y, \psi_x)$ is derived, $\zeta = v_x - u_y = \nabla^2\psi$ is the (vertical) vorticity, and β is a constant parameter (e.g. Pedlosky, 1987). (The same equation governs drift waves in plasma physics.) The nonlinear equation (2.1) has two conserved quadratic quantities (e.g. Ripa, 1981), the (pseudo)energy

$$\mathcal{E} = \frac{1}{2}\int |\nabla\psi|^2\,\mathrm{d}\mathbf{x},$$

and the pseudomomentum

$$\mathcal{P} = -\frac{1}{2\beta}\int \zeta^2\,\mathrm{d}\mathbf{x}.$$

(We use the terms pseudoenergy and pseudomomentum to designate the conserved quantities that are quadratic at leading order; these are often referred to as wave energy and wave momentum, and in many contexts they differ from what is usually termed energy and momentum (e.g., Grimshaw, 1984; Shepherd, 1990). The negative sign and β^{-1} factor in $\mathcal{P}$ are introduced for consistency with the general formulation of the next section.)

For small-amplitude motion, the nonlinear terms in (2.1) can be neglected. Solutions can then be sought in the form of plane waves,

$$\zeta = A\exp[\mathrm{i}(\mathbf{k}\cdot\mathbf{x} - \omega t)] + \text{c.c.}, \tag{2.2}$$

where the wavevector $\mathbf{k} = (k, l)$, the frequency ω and the amplitude A are constant, and where c.c. denotes the complex conjugate of the previous term. Introduction into (2.1) yields the dispersion relation

$$\omega = -\frac{\beta k}{k^2 + l^2} \tag{2.3}$$

which relates the frequency to the wavevector. Solution (2.2)–(2.3) describes waves with crests propagating in the direction of $\mathbf{k}$ at speed $c = \omega/|\mathbf{k}|$.

General solutions to the linearized vorticity equation can then be written as a superposition of plane waves in the form

$$\zeta = \sum_a A_a \exp[i(\mathbf{k}_a \cdot \mathbf{x} - \omega_a t)].$$

Here, $\sum_a$ denotes a sum over all admissible wavevectors $\mathbf{k}_a$. If the domain is periodic, say $[0, 2\pi] \times [0, 2\pi]$, the admissible k_a and l_a are integers; if the domain is unbounded, $\sum_a$ is understood as an integral over k_a and l_a. The reality of ζ is ensured by taking

$$A_a^* = A_{-a}, \tag{2.4}$$

where A_{-a} corresponds to the wavevector $-\mathbf{k}_a$. The conserved quadratic quantities can be written in terms of the amplitudes A_a as

$$\mathcal{E} = \frac{1}{2} \sum_a E_a |A_a|^2 \quad \text{and} \quad \mathcal{P} = \frac{1}{2} \sum_a P_a |A_a|^2, \tag{2.5}$$

where $E_a = (2\pi)^2/(k_a^2 + l_a^2)$ and $P_a = -(2\pi)^2/\beta$. We note that

$$\frac{E_a}{\omega_a} = \frac{P_a}{k_a} \tag{2.6}$$

is the wave action.

2.2 General linear waves

The construction outlined above for Rossby waves is just an instance of a construction that can be carried out for general systems governed by the system of differential equations

$$\mathbf{u}_t = \mathbf{L}\mathbf{u} + \mathbf{N}(\mathbf{u}), \tag{2.7}$$

where $\mathbf{u}$ denotes the n-dimensional dynamical variable, $\mathbf{L}$ is a (constant) matrix operator, and $\mathbf{N}(\mathbf{u})$ contains the nonlinear terms. We are interested in systems with at least two conserved quantities $\mathcal{E}$ and $\mathcal{P}$ that are quadratic at leading order in $\mathbf{u}$, i.e.,

$$\mathcal{E} = \mathcal{E}^{(2)} + \text{h.o.t.} \quad \text{and} \quad \mathcal{P} = \mathcal{P}^{(2)} + \text{h.o.t.},$$

where h.o.t. denotes terms that are cubic and higher order in $\mathbf{u}$, and

$$\mathcal{E}^{(2)} = \frac{1}{2} \int \mathbf{u}^\dagger \mathbf{E} \mathbf{u} \, dx \quad \text{and} \quad \mathcal{P}^{(2)} = \frac{1}{2} \int \mathbf{u}^\dagger \mathbf{P} \mathbf{u} \, dx, \tag{2.8}$$

where E and P are two self-adjoint operators, and † denotes transpose. (The form (2.7) with associated conserved quantities arises from the Hamiltonian nature of the conservative systems we are interested in; this implies, in particular, that L, E and P are related: $\mathsf{L} = \mathsf{JE}$ for some skew-adjoint operator J, and P is then such that $-\mathbf{u}_x = \mathsf{JP}\mathbf{u}$; cf. Morrison (1998); Vanneste and Shepherd (1999).)

When the nonlinear terms $\mathbf{N}(\mathbf{u})$ in (2.7) are neglected, plane-wave solutions of the form

$$\mathbf{u} = A\hat{\mathbf{u}}\exp[i(\mathbf{k}\cdot\mathbf{x} - \omega t)] + \text{c.c.}$$

can be introduced. The dispersion relation relating ω and $\mathbf{k}$ is then found by solving the matrix eigenvalue problem

$$-i\omega\hat{\mathbf{u}} = \hat{\mathsf{L}}\hat{\mathbf{u}}, \tag{2.9}$$

where $\hat{\mathsf{L}}$ denotes the operator L where the substitutions $\partial_x \mapsto ik$, $\partial_y \mapsto il$, etc. have been performed. In (2.9), ω is the eigenvalue and $\hat{\mathbf{u}}$ the eigenvector; the latter can be chosen up to an arbitrary factor. We will assume that there are n distinct real eigenvalues ω for each $\mathbf{k}$; thus, we consider spectrally stable systems, admitting no growing modes, and exclude polarized waves. The n eigenvalues correspond to distinct branches of the dispersion relation.

As in the case of Rossby waves, a superposition of plane waves of the form

$$\mathbf{u} = \sum_a A_a\hat{\mathbf{u}}_a\exp[i(\mathbf{k}_a\cdot\mathbf{x} - \omega_a t)] \tag{2.10}$$

can be considered. This time, the label a identifies the wavevector $\mathbf{k}_a$ and the branch of the dispersion relation, i.e. $a = (\mathbf{k}_a, p_a)$, $p_a = 1, 2, \cdots, n$. The eigenvalue problem (2.9) is such that if $(\omega_a, \hat{\mathbf{u}}_a)$ is a solution for the wavevector $\mathbf{k}_a$, $(\omega_{-a}, \hat{\mathbf{u}}_{-a}) = (-\omega_a, \hat{\mathbf{u}}_a^*)$ is a solution for the wavevector $-\mathbf{k}_a$. The reality of $\mathbf{u}$ is thus again ensured by requiring that (2.4) hold.

The quadratic quantities $\mathcal{E}^{(2)}$ and $\mathcal{P}^{(2)}$ are exactly conserved for the linearized dynamics. This is reflected in some orthogonality relations that the eigenvectors $\hat{\mathbf{u}}_a$ satisfy. To see this, we introduce (2.10) into (2.8) to find, for each $\mathbf{k}$,

$$\mathcal{E}^{(2)} = \frac{(2\pi)^d}{2}\sum_{p,q=1}^{n} A_p^* A_q \mathbf{u}_p^\dagger \mathsf{E}\mathbf{u}_q \exp[i(\omega_p - \omega_q)t],$$

where d is the number of spatial dimensions and † now denotes transposition and complex conjugation. A similar expression obtains for $\mathcal{P}^{(2)}$. Here, p and q label the branch of the dispersion relation whilst the dependence on $\mathbf{k}$ is not indicated for simplicity. The constancy of $\mathcal{E}^{(2)}$ and $\mathcal{P}^{(2)}$ immediately implies the following orthogonality relations:

$$(2\pi)^d\mathbf{u}_p^\dagger\mathsf{E}\mathbf{u}_q = E_p\delta_{p,q} \quad \text{and} \quad (2\pi)^d\mathbf{u}_p^\dagger\mathsf{P}\mathbf{u}_q = P_p\delta_{p,q} \qquad \text{for given } \mathbf{k}. \tag{2.11}$$

The constants E_p and P_p can be interpreted as the pseudoenergy and pseudomomentum of the p-th plane wave with wavevector $\mathbf{k}$. They depend on the normalization chosen for the eigenvectors $\hat{\mathbf{u}}$ and are related by $E_p/\omega_p = P_p/k_p$, so that (2.6) holds for our general formulation. The orthogonality relations (2.11) imply that the quadratic part of the pseudoenergy and pseudomomentum of the linear solutions (2.10) are just the sum of the contributions of each wave mode; that is, analogues of (2.5) hold in general for $\mathcal{E}^{(2)}$ and $\mathcal{P}^{(2)}$.

Remark. We have here considered only one conserved pseudomomentum $\mathcal{P}$ in addition to the pseudoenergy. For some systems, such as surface gravity waves, the pseudomomentum is a vector whose components, say $\mathcal{P}$ and $\mathcal{P}'$, (or more) can be conserved; they are connected by the relationship $P_a/k_a = P'_a/l_a$.

Internal gravity waves. Internal gravity waves, propagating in stratified fluids, can be described using the Boussinesq approximation. In two dimensions, with (x, z) as horizontal and vertical coordinates, respectively, the corresponding equations of motion for the vorticity ζ and the buoyancy $b = -g\rho/\rho_0$ (with $\rho + \rho_0$ the density whose constant mean is $\rho_0 \gg \rho$) read

$$\zeta_t - b_x + \psi_x\zeta_z - \psi_z\zeta_x = 0, \quad b_t + N^2\psi_x + \psi_x b_z - \psi_z b_x = 0, \tag{2.12}$$

where $\nabla^2\psi = \zeta$ and $N > 0$ is the (constant) Brunt–Väisälä frequency (cf., e.g., Gill, 1982). These equations can be cast in the general form (2.7) with $\mathbf{u} = (\zeta, b)$

$$\mathsf{L} = \begin{pmatrix} 0 & \partial_x \\ -N^2\partial_x\nabla^{-2} & 0 \end{pmatrix}, \quad \mathsf{E} = \begin{pmatrix} -\nabla^{-2} & 0 \\ 0 & N^{-2} \end{pmatrix}, \quad \mathsf{P} = -\begin{pmatrix} 0 & N^{-2} \\ N^{-2} & 0 \end{pmatrix}.$$
$$\tag{2.13}$$

The eigenvalue problem (2.9) takes the form

$$\begin{pmatrix} \omega & k \\ N^2k/K^2 & \omega \end{pmatrix}\hat{\mathbf{u}} = 0,$$

where we have written the wavenumber as $\mathbf{k} = (k, m)$ and $K = |\mathbf{k}| = (k^2 + m^2)^{1/2}$. The two branches of solutions, with

$$\omega_p = (-1)^p\frac{Nk}{K} \quad \text{and} \quad \hat{\mathbf{u}}_p = \begin{pmatrix} K \\ (-1)^{p+1}N \end{pmatrix}, \quad \text{for} \quad p = 1, 2,$$

correspond to waves propagating in opposite directions. The orthogonality relations (2.11) can be verified explicitly; with the normalization chosen, the pseudoenergy and pseudomomentum are found to be

$$E_p = 8\pi^2 \quad \text{and} \quad P_p = 8\pi^2\frac{k}{\omega_p}$$

and obviously satisfy (2.6).

3 Interaction equations

The superposition of waves (2.10) provides the general solution for the linearized equations of motion. It remains useful when the nonlinear terms are taken into account. In this case, the wave amplitudes A_a do not remain constant: the nonlinear terms lead to energy transfers between different waves which change their amplitudes. We now derive the evolution equations governing these changes.

3.1 Derivation

For Rossby waves, we introduce the expansion

$$\zeta = \sum_a A_a(t) \exp[i(\mathbf{k}_a \cdot \mathbf{x} - \omega_a t)] \tag{3.1}$$

into (2.1). Taking the dispersion relation into account, and simplifying the exponentials, we obtain

$$\dot{A}_a = \frac{1}{2} \sum_{bc} I_a^{bc} A_b^* A_c^* \exp(2i\Omega_{abc}t)\delta_{\mathbf{k}_a+\mathbf{k}_b+\mathbf{k}_c}, \tag{3.2}$$

where $\Omega_{abc} = (\omega_a + \omega_b + \omega_c)/2$, and the interaction coefficients

$$I_a^{bc} = (k_b l_c - k_c l_b)\left(\frac{1}{k_c^2 + l_c^2} - \frac{1}{k_b^2 + l_b^2}\right) \tag{3.3}$$

have been symmetrized (so that $I_a^{bc} = I_a^{cb}$) for convenience (Ripa, 1981).

In (3.2), the sum involves only modes b and c satisfying the interaction condition

$$\mathbf{k}_a + \mathbf{k}_b + \mathbf{k}_c = 0. \tag{3.4}$$

This implies that $\delta_{\mathbf{k}_a+\mathbf{k}_b+\mathbf{k}_c}$ should be interpreted as the product of Kronecker symbols,

$$\delta_{\mathbf{k}_a+\mathbf{k}_b+\mathbf{k}_c} = \delta_{k_a+k_b+k_c,0}\, \delta_{l_a+l_b+l_c,0},$$

for periodic domains, and as the product of delta functions

$$\delta_{\mathbf{k}_a+\mathbf{k}_b+\mathbf{k}_c} = \delta(k_a + k_b + k_c)\delta(l_a + l_b + l_c),$$

in unbounded domains. Note that only + signs appear in (3.4), when some authors have ± signs: this is the result of considering both wavevectors $\mathbf{k}_a$ and $-\mathbf{k}_a$ in the expansion (3.1).

The interaction equations (3.2) are in fact obtained for general wave systems, with the proviso that additional terms, cubic and higher order, are included when the nonlinearity of the governing equations is not exactly quadratic. To obtain interaction equations in the general case, we introduce the expansion

$$\mathbf{u} = \sum_a A_a(t)\hat{\mathbf{u}}_a \exp[i(\mathbf{k}_a \cdot \mathbf{x} - \omega_a t)] \tag{3.5}$$

into (2.7). Using (2.9) and simplifying the exponentials does not directly give equations for each of the A_a, since modes with the same wavevector but corresponding to different branches of the dispersion relation remain coupled. To isolate them, the orthogonality relations (2.11) must be used. Multiplying the equations by $\hat{\mathbf{u}}_a^\dagger \mathsf{E}$, with $a = (\mathbf{k}_a, p_a)$ and using (2.11)$_1$ gives the interaction equations in the form

$$\dot{A}_a = \frac{1}{2} \sum_{bc} I_a^{bc} A_b^* A_c^* \exp(2i\Omega_{abc}t)\delta_{\mathbf{k}_a+\mathbf{k}_b+\mathbf{k}_c} + \text{h.o.t.}, \tag{3.6}$$

with

$$I_a^{bc} = (2\pi)^d \hat{\mathbf{u}}_a^\dagger \mathsf{E} \left[\hat{\mathbf{N}}^{(2)}(\hat{\mathbf{u}}_b, \hat{\mathbf{u}}_c) + \hat{\mathbf{N}}^{(2)}(\hat{\mathbf{u}}_c, \hat{\mathbf{u}}_b) \right]^* / E_a, \tag{3.7}$$

and $\hat{\mathbf{N}}^{(2)}$ denotes the quadratic part of the nonlinear terms, with $\partial_x \mapsto ik$, etc.

Internal gravity waves. The nonlinear terms in the equations (2.12) for internal gravity waves are

$$\mathbf{N}(\mathbf{u}) = \mathbf{N}^{(2)}(\mathbf{u}) = \begin{pmatrix} -\psi_x \zeta_z + \psi_z \zeta_x \\ -\psi_x b_z + \psi_z b_x \end{pmatrix}.$$

We can compute the corresponding interaction coefficients using (3.7). We find

$$\hat{\mathbf{N}}^{(2)}(\hat{\mathbf{u}}_b, \hat{\mathbf{u}}_c) = -(k_b m_c - k_c m_b) \begin{pmatrix} K_c/K_b \\ (-1)^{p_c+1} N/K_b \end{pmatrix}$$

and, after a short calculation,

$$I_a^{bc} = \frac{k_c m_b - k_b m_c}{2K_a K_b K_c} \left[K_c(K_c + (-1)^{p_a+p_c} K_a) - K_b(K_b + (-1)^{p_a+p_b} K_a) \right].$$

3.2 Properties of the interaction coefficients

The conservation of $\mathcal{E}$ and $\mathcal{P}$ imposes constraints on the form of the interaction coefficients I_a^{bc} which we now discuss.

Consider first the pseudoenergy which can be written in terms of the wave amplitudes $A_a(t)$ as

$$\mathcal{E} = \mathcal{E}^{(2)} + \mathcal{E}^{(3)} + \cdots = \frac{1}{2} \sum_a E_a |A_a|^2 + \frac{1}{6} \sum_{abc} S_{abc} A_a^* A_b^* A_c^* \exp(2i\Omega_{abc} t) + \cdots,$$

for some coefficients S_{abc}, with $\cdots$ denoting terms of order higher than 3. Now, from (3.6) we find that

$$\dot{\mathcal{E}}^{(2)} = \frac{1}{2} \sum_{abc} E_a \left(I_a^{bc} A_a^* A_b^* A_c^* \exp(-2i\Omega_{abc} t) + \text{c.c.} \right) + \cdots$$

and

$$\dot{\mathcal{E}}^{(3)} = \frac{i}{3} \sum_{abc} \Omega_a S_{abc} A_a^* A_b^* A_c^* \exp(2i\Omega_{abc} t)] + \cdots.$$

The vanishing of $\dot{\mathcal{E}}$ then imposes at leading order that

$$E_a I_a^{bc} + E_b I_b^{ca} + E_c I_c^{ab} = -iS_{abc}\Omega_{abc}.$$

Similarly, the vanishing of $\dot{\mathcal{P}}$ imposes that

$$P_a I_a^{bc} + P_b I_b^{ca} + P_c I_c^{ab} = -iT_{abc}\Omega_{abc},$$

where the T_{abc} are the coefficients of $A_a A_b A_c$ in the expansion of $\mathcal{P}^{(3)}$ (Ripa, 1981; Vanneste and Vial, 1994). These two relations are useful if $S_{abc}\Omega_{abc} = T_{abc}\Omega_{abc} = 0$. This happens in two important instances: (i) when the pseudoenergy $\mathcal{E}$ and pseudomomentum $\mathcal{P}$ are exactly quadratic (as is the case for the Rossby waves and internal gravity waves encountered so far); or (ii) when the three waves a, b and c satisfy the resonance condition

$$\omega_a + \omega_b + \omega_c = 0 \tag{3.8}$$

(hence $\Omega_{abc} = 0$) in addition to the interaction condition (3.4). Case (ii) will be seen below to be crucial for weakly nonlinear interactions. Assuming either (i) or (ii), it is easy to derive the relations

$$\frac{E_a I_a^{bc}}{s_b - s_c} = \frac{E_b I_b^{ca}}{s_c - s_a} = \frac{E_c I_c^{ab}}{s_a - s_b} \quad \text{and} \quad \frac{P_a I_a^{bc}}{c_b - c_c} = \frac{P_b I_b^{ca}}{c_c - c_a} = \frac{P_c I_c^{ab}}{c_a - c_b}, \tag{3.9}$$

where $s_a = 1/c_a = k_a/\omega_a$ is the slowness of wave a in the x-direction. If (3.8) holds, i.e., for resonant waves, the more useful forms

$$\frac{E_a I_a^{bc}}{\omega_a} = \frac{E_b I_b^{ca}}{\omega_b} = \frac{E_c I_c^{ab}}{\omega_c} \quad \text{and} \quad \frac{P_a I_a^{bc}}{k_a} = \frac{P_b I_b^{ca}}{k_b} = \frac{P_c I_c^{ab}}{k_c} \tag{3.10}$$

follow. We stress that the properties (3.9)-(3.10) result from the conservation laws for $\mathcal{E}$ and $\mathcal{P}$; they can also be established from the explicit form of the interaction coefficients I_a^{bc}, but this often requires tedious computations, unless the Hamiltonian or Lagrangian structure of the equations of motion is used explicitly (see section 7).

Remark. Interaction equations similar to (3.6) can also be derived for systems without particular conservation laws such as systems with dissipation. In this case, the orthogonality relations (2.11) that need to be used have the form $(2\pi)^d \mathbf{u}_p^+ \mathbf{u}_q = E_p \delta_{p,q}$, where $\mathbf{u}^+$ is the left eigenvector of L.

4 Triad interactions

So far, no approximations have been made: (3.6) (if the higher-order terms were computed) is simply a reformulation of the general system (2.7) which uses the amplitudes $A_a(t)$ rather than the original $\mathbf{u}(\mathbf{x}, t)$ as dependent variables. In what follows, we are interested in the weakly nonlinear regime which assumes that the wave amplitudes are small: $|A_a| = O(\epsilon)$, with $\epsilon \ll 1$. We also consider situations in which only a few waves are initially excited; for a finite time (typically scaling like ϵ^{-1}), the truncation of the interaction equations (3.6) to retain only those amplitudes that are excited initially can then be justified using a multiple-scale approach (Bretherton, 1964). When the lowest-order nonlinearity is quadratic, the smallest set of waves that can be considered is a set of three waves — a wave triad.

4.1 Resonant triads

Let us consider the amplitude equations for a triad of waves a, b and c satisfying the interaction condition (3.2). Neglecting cubic and higher-order terms, their amplitudes are governed by the truncation of (3.6), namely

$$\begin{aligned}
\dot{A}_a &= I_a^{bc} A_b^* A_c^* \exp(2i\Omega_{abc}t), \\
\dot{A}_b &= I_b^{ca} A_c^* A_a^* \exp(2i\Omega_{abc}t), \\
\dot{A}_c &= I_c^{ca} A_a^* A_b^* \exp(2i\Omega_{abc}t).
\end{aligned} \tag{4.1}$$

It is only when the triad is near resonant, more precisely when

$$\omega_a + \omega_b + \omega_c = 2\Omega_{abc} = O(\epsilon), \tag{4.2}$$

that the quadratic interactions are significant. This can be seen, for instance, by noting that the change of variables from A_a to B_a, with

$$A_a = B_a - \frac{iI_a^{bc}}{2\Omega_{abc}} B_b^* B_c^* \exp(2i\Omega_{abc}t), \tag{4.3}$$

and similar definitions of B_c and B_b, leads to

$$\dot{B}_a = \dot{B}_b = \dot{B}_c = O(\epsilon^3).$$

Thus the transformation (4.3) eliminates the quadratic nonlinearities from (4.1); B_a, B_b and B_c, and hence A_a, A_b and A_c, change only by a small $O(\epsilon)$ amount for $t = O(\epsilon^{-1})$. The variable transformation cannot be used, however, when (4.2) holds, and A_a, A_b and A_c change by an $O(1)$ amount over a $O(\epsilon^{-1})$ time scale. (Note that a triad cannot be isolated if one of the waves and the complex conjugate of another (say with wavenumber and frequency $-k_c$ and $-\omega_c$) form a resonant triad with a fourth wave; when this is the case, a system of four coupled amplitude equations must be considered.)

The interaction and resonance conditions

$$\mathbf{k}_a + \mathbf{k}_b + \mathbf{k}_c = 0 \quad \text{and} \quad \omega_a + \omega_b + \omega_c = 0$$

make up a system of $d+1$ algebraic equations for the $3d$ unknowns that are the wavenumbers. In periodic domains these wavenumbers can only have discrete values, and the interaction and resonance conditions can only be satisfied exceptionally. In unbounded domains, however, the wavenumbers are continuous, and resonances abound.

Resonant triads are usually found using numerical calculations. They can also be found using a graphical method (Simmons, 1969) which we now demonstrate. The idea is to plot the dispersion relations $\omega_a = \omega(k_a, l_a)$ and $\omega_c = \omega(k_c, l_c)$ for two fixed wavenumbers l_a and l_c on one graph, and the dispersion relation $\omega_b = \omega(k_b, l_b)$ for fixed l_b with $l_a + l_b + l_c = 0$ on another graph. Sliding the origin of the second graph along the curve $\omega_a = \omega(k_a, l_a)$, leads to resonant values of k_a, k_b and k_c whenever the curve $\omega_b = \omega(k_b, l_b)$ on the second graph intersects the $\omega_c = \omega(k_c, l_c)$ on the first graph. See Figure 1 for an example of application to Rossby waves.

4.2 Solution of the triad equations

Assuming (4.2), the interaction coefficients can be modified by small corrections introducing negligible, $O(\epsilon^3)$, errors in (4.1) so that (3.10) hold. The simple scaling

$$A_a = \epsilon |I_b^{ca} I_c^{ab}|^{-1/2} \alpha_a, \quad A_b = \epsilon |I_c^{ab} I_a^{bc}|^{-1/2} \alpha_b, \quad A_c = \epsilon |I_a^{bc} I_b^{ca}|^{-1/2} \alpha_c$$

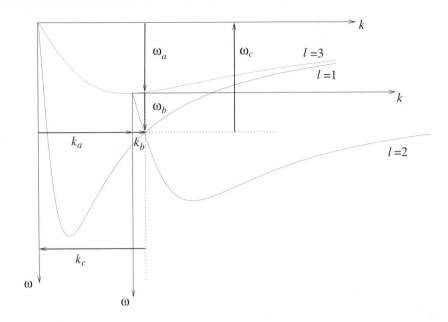

Figure 1. Construction of a resonant triad for Rossby waves. The dispersion relations for $l_a = 3$ and $l_c = 1$ are plotted on one graph, the dispersion relation for $l_b = 2$ on the other.

and the rescaling of t by ϵ then turns (4.1) into the universal form

$$\begin{aligned}
\dot{\alpha}_a &= \sigma_a \alpha_b^* \alpha_c^* \exp(2\mathrm{i}\Omega t) \\
\dot{\alpha}_b &= \sigma_b \alpha_c^* \alpha_a^* \exp(2\mathrm{i}\Omega t) \\
\dot{\alpha}_c &= \sigma_c \alpha_a^* \alpha_b^* \exp(2\mathrm{i}\Omega t),
\end{aligned} \tag{4.4}$$

where

$$\Omega = \Omega_{abc}/\epsilon = O(1), \quad \sigma_a = \mathrm{sign}\, I_a^{bc}, \quad \sigma_b = \mathrm{sign}\, I_b^{ca} \quad \text{and} \quad \sigma_c = \mathrm{sign}\, I_c^{ab}$$

(e.g. Craik, 1985). The qualitative behaviour can be inferred from the Manley–Rowe relations

$$\frac{\mathrm{d}}{\mathrm{d}t}\left(\sigma_a|\alpha_a|^2 - \sigma_b|\alpha_b|^2\right) = \frac{\mathrm{d}}{\mathrm{d}t}\left(\sigma_b|\alpha_b|^2 - \sigma_c|\alpha_c|^2\right) = \frac{\mathrm{d}}{\mathrm{d}t}\left(\sigma_c|\alpha_c|^2 - \sigma_a|\alpha_a|^2\right) = 0 \tag{4.5}$$

which are immediately derived from (4.4).

There are only two cases to consider: (i) one of the interaction coefficients has a different sign from the other two, or (ii) the three interactions coefficients have the same sign. Because of (3.10), the sign of the wave actions $E_a/\omega_a = P_a/k_a$ can be considered

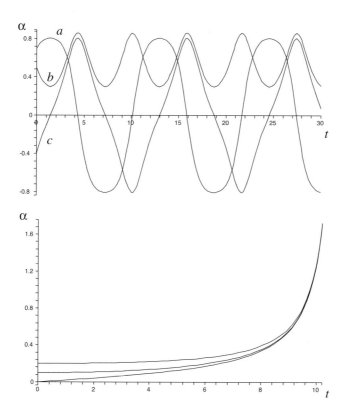

Figure 2. Solutions of the triad equations (4.4) as a function of t. The top panel has been obtained for the sign combination (i) in (4.6), with $\alpha_a(0) = 0.7$, $\alpha_b(0) = 0.5$ and $\alpha_c(0) = -0.4$; the bottom panel has been obtained for the sign combination (ii), with $\alpha_a(0) = 0.1$, $\alpha_b(0) = 0.2$ and $\alpha_c(0) = 0$. The frequencies satisfy $\Omega = 0$; since the amplitude are real initially, they remain so for all time.

in place of that of the interaction coefficients I_a^{bc}. We can in fact restrict our attention to two sign combinations:

$$(i) \quad -\sigma_a = \sigma_b = \sigma_c = 1 \quad \text{or} \quad (ii) \quad \sigma_a = \sigma_b = \sigma_c = 1. \tag{4.6}$$

Other sign combinations reduce to either of (4.6) by changes of signs of the α_a. Case (i) is the most common: it arises when all the wave pseudoenergies (or pseudomomenta) have the same sign (since the near resonance implies that the sign of one of the frequencies (or wavenumbers) differs from the sign of the other two). The Manley–Rowe relations (4.5) then indicate that the three amplitudes are bounded by the maximum of the initial amplitudes. As shown below, the three waves simply exchange energy in a manner that turns out to be periodic, as illustrated in the top panel of Figure 2.

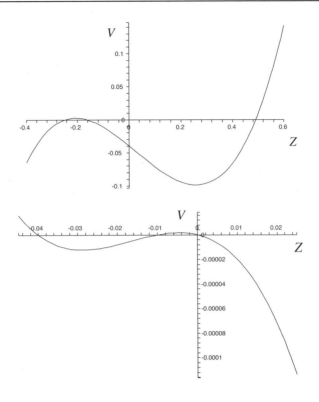

Figure 3. Potential $V(Z)$ defined in (4.9) for the parameters of the two panels of Figure 2. The qualitative behaviour of $Z(t)$ is deduced by thinking of $Z(t)$ as the coordinate of a zero-energy particle in the potential $V(Z)$.

Case (ii) can occur when the wave pseudoenergy is not sign definite: if the wave with the largest frequency (in absolute value) has a pseudoenergy that has opposite sign to the other two waves, then the actions E_a/ω_a, E_b/ω_b and E_c/ω_c have the same sign, and (ii) holds. The Manley–Rowe relations impose no bounds on the wave amplitudes; these can grow simultaneously, in fact grow so rapidly that they reach a singularity in a finite time as shown in the bottom panel of Figure 2. Since this occurs from arbitrarily small initial amplitudes (for resonant triads), this provides an interesting mechanism of instability, called explosive resonant interaction, that is nonlinear in nature. Obviously this cannot happen for waves propagating in resting fluids (these have positive pseudoenergy), but this is possible when the fluid is in motion (see, e.g., Craik and Adam, 1979; Vanneste, 1995, for examples).

The triad equations (4.4) can in fact be solved in closed form. To see this, define

$$Z(t) = \sigma_a \left[|\alpha_a(t)|^2 - |\alpha_a(0)|^2\right] = \sigma_b \left[|\alpha_b(t)|^2 - |\alpha_b(0)|^2\right] = \sigma_c \left[|\alpha_c(t)|^2 - |\alpha_c(0)|^2\right],$$ (4.7)

where the equalities follow from (4.5), and note that

$$W = |\alpha_a \alpha_b \alpha_c| \sin\phi + \Omega Z(t),$$

where

$$\phi = \arg\alpha_a + \arg\alpha_b + \arg\alpha_c - 2\Omega t,$$

is a constant (Bretherton, 1964). Using this, one can derive from (4.4) the single equation

$$\frac{1}{2}\dot{Z}^2 + V(Z) = 0,$$ (4.8)

where the cubic

$$V(Z) = 2\left[(W - \Omega Z)^2 - (\sigma_a Z + |\alpha_a(0)|^2)(\sigma_b Z + |\alpha_b(0)|^2)(\sigma_c Z + |\alpha_c(0)|^2)\right]$$ (4.9)

can be interpreted as a potential. The qualitative behaviour of the solution is then readily deduced from the graph of $V(Z)$, thinking of $Z(t)$ (with $Z(0) = 0$) as the position of a particle with zero energy in the potential $V(Z)$ (cf. Craik, 1985). This is illustrated in Figure 3. The top panel shows $V(Z)$ for the parameters of the top panel of Figure 2 (these give $W = 0$). Clearly, $Z(t)$ can be predicted to oscillate periodically between $|\alpha_a(0)|^2$ and $-|\alpha_c(0)|^2$. This corresponds to waves a and c exchanging their energy completely, while the amplitude of wave b only fluctuates between two non-zero values; this is consistent with Figure 2. A nonzero value for the constant W, determined by the initial phase of the waves, shifts the potential $V(z)$ upward, thus inhibiting the energy exchanges. Similarly, a non-zero Ω, corresponding to near rather than exact resonance, inhibits the energy exchanges between the three waves.

The bottom panel of Figure 3 displays the potential $V(Z)$ for the explosive interaction (case (ii)) with the parameters corresponding to the bottom panel of Figure 2. It is clear from the form of the potential that $Z(t)$, and hence the wave amplitudes, grow unboundedly as observed. Note that a non-zero W, while it slows down the growth, does not suppress it; in contrast, a non-zero Ω can lead to oscillations also in case (ii) provided that the initial wave amplitudes are small enough.

Closed-form expressions for $Z(t)$ can be derived from (4.8)–(4.9) in terms of elliptic functions whose parameters depend on the zeroes of $V(Z)$. The moduli of the wave amplitudes $|\alpha_a(t)|$, $|\alpha_b(t)|$ and $|\alpha_c(t)|$ then follow from (4.7). The phases $\arg\alpha_a(t)$, $\arg\alpha_b(t)$ and $\arg\alpha_c(t)$ can finally be reconstructed by quadrature from

$$\frac{\mathrm{d}}{\mathrm{d}t}\arg\alpha_a = -\sigma_a \frac{|\alpha_b||\alpha_c|}{|\alpha_a|}\sin\phi,$$

and similar equations for $\arg\alpha_b$ and $\arg\alpha_c$. Note that, unlike the amplitude moduli, the phases are generally not periodic in t; hence the complex amplitudes α_a, α_b and α_c are not periodic (see Alber et al., 1998).

4.3   Wave instability

A straightforward yet very useful application of the interaction equations (4.4) is the study of the stability of small-amplitude waves. Here, we consider a medium at rest, with waves of sign-definite pseudoenergy (case (i) in (4.6)). The medium is stable, but the plane waves can be unstable and disintegrate when disturbed by small perturbations. The discussion of wave stability is clean-cut when the plane waves are exact solutions of the fully nonlinear equations of motion; it can be verified that this is the case both for the Rossby and internal gravity waves discussed in these lectures.

The stability of finite-amplitude waves can be studied in the standard way, by considering the linearized evolution equations for a small perturbation to the waves. Because of the temporal and spatial periodicity of the waves, this evolution equation has periodic coefficients, and Floquet theory (e.g Ince, 1956) must be employed. This is a difficult task generally requiring numerical computations. Some information can be gained, however, by considering the stability of small-amplitude waves, and by assuming that their perturbation consists of two plane waves. The significant energy exchange necessary for instability is possible only if the three waves form a near-resonant triad.

Let a be the wave whose stability we consider, often called the primary wave, and b and c be the so-called secondary waves, i.e. the two waves constituting the perturbation. The assumptions are that the wave amplitudes satisfy

$$|\alpha_b|, |\alpha_c| \ll |\alpha_a| \ll 1.$$

With these, the interactions equations (4.4) simplify: α_a can be considered as a constant, and α_b and α_c obey the linear system

$$\dot{\alpha}_b = \alpha_a^* \alpha_c^* \exp(2i\Omega t), \quad \dot{\alpha}_c = \alpha_a^* \alpha_b^* \exp(2i\Omega t).$$

This approximation is often referred to as "pump-wave approximation" and can also be relevant in situations in which the amplitude of wave a is kept constant artificially.

Substituting solutions of the form

$$\alpha_b = \exp[(\lambda + i\Omega)t]\hat{\alpha}_b \quad \text{and} \quad \alpha_c = \exp[(\lambda^* + i\Omega)t]\hat{\alpha}_c,$$

where $\hat{\alpha}_b$, $\hat{\alpha}_c$ and λ are complex constant, leads to

$$\lambda = \pm \left[|\alpha_a|^2 - \Omega^2\right]^{1/2},$$

or, returning to unscaled amplitude and time,

$$\lambda = \pm \left[I_b^{ca} I_c^{ab} |A_a|^2 - \Omega_{abc}^2\right]^{1/2}.$$

Clearly, a (real) positive value of λ indicates the instability of wave a: it occurs for any amplitude $|A_a|$ if the triad a, b and c is resonant, and provided that $|A_a| > \Omega_{abc}/(I_b^{ca} I_c^{ab})^{1/2}$ if it is not. The sign combination chosen in (i) corresponds to the primary wave a having the largest frequency (in absolute value) of the triad. This leads to Hasselmann (1967)'s criterion for the instability of waves: a small-amplitude wave is unstable if it is the highest-frequency member of a resonant triad. Given the abundance of resonant triads, this means that most waves are unstable, although the growth rate of the instabilities is small, since it is proportional to the primary-wave amplitude.

It is interesting to consider situations in which the secondary waves have asymptotically large wavenumbers, i.e. $|\mathbf{k}_b|, |\mathbf{k}_c| \gg |\mathbf{k}_a|$. The interaction condition then imposes that $\mathbf{k}_b \approx -\mathbf{k}_c$. If waves b and c belong to the same branch of the dispersion relation, $\omega_b \approx -\omega_c$; the relation $\omega_c = \omega(k_c) = \omega(-k_b - k_a)$ can then be expanded in the resonance condition (3.8) to find

$$\omega(\mathbf{k}_a) = \mathbf{k}_a \cdot \frac{\partial \omega}{\partial \mathbf{k}}(\mathbf{k}_b).$$

Thus instability is possible for short secondary waves, provided that their group velocity $\partial \omega / \partial \mathbf{k}$ satisfy this condition; for one-dimensional waves or more generally if the group velocity is in the direction of $\mathbf{k}_a$, this imposes that this group velocity of the secondary waves match the phase velocity of the primary wave.

For internal gravity waves, the two secondary waves can belong to the two different branches of the dispersion relation which differ by sign. Thus, $\mathbf{k}_b \approx -\mathbf{k}_c$ implies $\omega_b \approx \omega_c \approx -\omega_a/2$. This type of instability turns out to lead to the largest growth rates when the frequency ω_a is sufficiently large.

5 Interactions of wavepackets

So far, we have examined the interaction of plane waves which, in unbounded domains, have infinite energy. There is considerable interest in examining the more realistic situation of interacting wavepackets, which are localized in space. Such wavepackets are obtained by superposition of plane waves, with amplitudes that are narrowly peaked around a central wavevector, for instance

$$A(\mathbf{k}, t) \propto \exp(-|\mathbf{k} - \mathbf{k}_a|^2/\delta^2), \quad \text{with } \delta \ll 1.$$

Wavepackets are best described in terms of the space- and time-dependent amplitude $B(\mathbf{x}, t)$ of the wavepacket envelope. For a wavepacket with central wavevector $\mathbf{k}_a$, this amplitude is related to the plane-wave amplitude according to

$$B_a(\mathbf{x}, t) = \int A(\mathbf{k}_{a'}, t) \exp\left[i(\mathbf{k}_{a'} - \mathbf{k}_a) \cdot \mathbf{x} - (\omega(\mathbf{k}_{a'}) - \omega_a)t\right] d\mathbf{k}_{a'}. \quad (5.1)$$

If $A(\mathbf{k}, t)$ is localized over a region of size $\delta \ll 1$ around $\mathbf{k}_a$, B_a extends over a region of size $\delta^{-1} \gg 1$ in space.

We are interested in the nonlinear interactions of three wavepackets, with respective central wavenumbers $\mathbf{k}_a$, $\mathbf{k}_b$ and $\mathbf{k}_c$ satisfying the interaction condition (3.2). We now derive evolution equations for the corresponding amplitudes $B_a(\mathbf{x}, t)$, $B_b(\mathbf{x}, t)$ and $B_c(\mathbf{x}, t)$ that govern these interactions. We first note from (5.1) that

$$\partial_t B_a(\mathbf{x}, t) = \int \left[\dot{A}_{a'} - iA_{a'}(\omega_{a'} - \omega_a)\right] \exp\left[i(\mathbf{k}_{a'} - \mathbf{k}_a) \cdot \mathbf{x} - (\omega_{a'} - \omega_a)t\right] d\mathbf{k}_{a'}. \quad (5.2)$$

Taylor expanding $\omega_{a'} - \omega_a \approx \mathbf{c}_a \cdot (\mathbf{k}_{a'} - \mathbf{k}_a)$, where

$$\mathbf{c}_a = \frac{\partial \omega}{\partial \mathbf{k}}(\mathbf{k}_a)$$

is the group velocity of wave a, we have

$$\int i A_{a'} (\omega_{a'} - \omega_a) \exp\left[i(\mathbf{k}_{a'} - \mathbf{k}_a) \cdot \mathbf{x} - (\omega_{a'} - \omega_a)t\right] d\mathbf{k}_a'$$
$$\approx \int i A_{a'} \mathbf{c}_a \cdot \nabla \exp\left[i(\mathbf{k}_{a'} - \mathbf{k}_a) \cdot \mathbf{x} - (\omega_{a'} - \omega_a)t\right] d\mathbf{k}_{a'} = \mathbf{c}_a \cdot \nabla B_a(\mathbf{x}, t).$$

On the other hand, neglecting cubic terms, the interaction equations (3.6) give

$$\int \dot{A}_{a'} \exp\left[i(\mathbf{k}_{a'} - \mathbf{k}_a) \cdot \mathbf{x} - (\omega_{a'} - \omega_a)t\right] d\mathbf{k}_a'$$
$$\approx \iiint I_{a'}^{b'c'} A_{b'}^* A_{c'}^* \delta(\mathbf{k}_{a'} + \mathbf{k}_{b'} + \mathbf{k}_{c'}) \exp\left[i(\mathbf{k}_{a'} - \mathbf{k}_a) \cdot \mathbf{x} + (\omega_a + \omega_{b'} + \omega_{c'})t\right] d\mathbf{k}_{a'} d\mathbf{k}_{b'} d\mathbf{k}_{c'}$$
$$\approx I_a^{bc} B_b^*(\mathbf{x}, t) B_c^*(\mathbf{x}, t) \exp(2i\Omega_{abc}t),$$

where (3.2) and the approximation $I_{a'}^{b'c'} \approx I_a^{bc}$ have been used to derive the last equality. Introducing these results into (5.1) and similar expressions for the amplitudes of wavepackets b and c lead to the interaction equations

$$\begin{aligned}
\dot{B}_a + \mathbf{c}_a \cdot \nabla B_a &= I_a^{bc} B_b^* B_c^* \exp(2i\Omega_{abc}t), \\
\dot{B}_b + \mathbf{c}_b \cdot \nabla B_b &= I_b^{ca} B_c^* B_a^* \exp(2i\Omega_{abc}t), \\
\dot{B}_c + \mathbf{c}_c \cdot \nabla B_c &= I_c^{ab} B_a^* B_b^* \exp(2i\Omega_{abc}t).
\end{aligned}$$

These are similar to the interaction equations for plane waves (4.4), with the crucial difference that they are partial rather than ordinary differential equations, with the presence of the advective terms $\mathbf{c} \cdot \nabla$ on the left-hand sides. Note that this term is of a similar order of magnitude as the nonlinear terms if $\delta = O(\epsilon)$. Assuming this, the transformation $B_a = \epsilon |I_b^{ca} I_c^{ab}|^{1/2} \beta_a$, etc., and the rescaling of t and $\mathbf{x}$ by ϵ gives the universal form

$$\begin{aligned}
\dot{\beta}_a + \mathbf{c}_a \cdot \nabla \beta_a &= \sigma_a \beta_b^* \beta_c^*, \\
\dot{\beta}_b + \mathbf{c}_b \cdot \nabla \beta_b &= \sigma_b \beta_c^* \beta_a^*, \\
\dot{\beta}_c + \mathbf{c}_c \cdot \nabla \beta_c &= \sigma_c \beta_a^* \beta_b^*,
\end{aligned} \qquad (5.3)$$

where we have assumed resonance $\Omega_{abc} = 0$ for simplicity. As before only the two cases (4.6) for the values of σ_a, σ_b and σ_c need to be considered.

Remarkably, the system (5.3) in one spatial dimension is integrable by inverse scattering like, for instance, the Korteweg–de Vries equation (cf. Chapter 1). The solutions for each of the envelopes, assumed to be initially well separated, then consists in a number of solitons and a radiation part, which because of the lack of dispersion in (5.3) does not decay. This is reviewed in details by Kaup et al. (1979). The interactions between wavepackets depend crucially on the relative group velocities as well as on the signs σ_a, σ_b and σ_c. For instance, for the sign combination (i) in (4.6), and assuming that $|c_b| < |c_a| < |c_c|$, wavepacket a loses its solitons to wavepackets b and c in a process known as soliton decay. In case (ii), explosive interaction, with finite-time singularity, is possible provided that the wavepacket with intermediate group velocity covers a sufficient area and contains solitons.

In addition to the inverse-scattering solutions, several explicit solutions of (5.3) can be found, as travelling waves, for instance, or by making a pump-wave approximation (see Craik, 1985, for references).

6 Quartet interactions

Resonant or near-resonant interactions are the only interactions that affect wave amplitudes to a significant extent. When three-wave resonances are impossible because of the nature of the dispersion relation, then the four-wave resonances that may be associated with cubic nonlinearity control the evolution of the wave amplitudes. This is (famously) the case for surface gravity waves for which the dispersion relation, namely

$$\omega = \pm(g|\mathbf{k}|)^{1/2}, \tag{6.1}$$

does not allow three-wave resonance (e.g. Phillips, 1977). (Obviously, four-wave resonances are also crucial for evolution equations whose nonlinearity is cubic at lowest order.)

A quartet of waves a, b, c and d interacts provided that the wavevectors satisfy the interaction condition

$$\mathbf{k}_a + \mathbf{k}_b + \mathbf{k}_c + \mathbf{k}_d = 0. \tag{6.2}$$

The corresponding resonance condition,

$$\omega_a + \omega_b + \omega_c + \omega_d = 0, \tag{6.3}$$

must be satisfied approximately, in fact with an $O(\epsilon^2)$ error, for the interaction to be significant.

For surface waves, resonant quartets can be visualized on a simple graph: since the dispersion relation (6.1) depends only on $\mathbf{k}$ in a scale-invariant fashion, we can restrict our attention to quartets with

$$\mathbf{k}_a + \mathbf{k}_b = -\mathbf{k}_c - \mathbf{k}_d = \begin{pmatrix} 1 \\ 0 \end{pmatrix},$$

the other quartets being obtained by scaling and rotation. The resonance condition reads

$$|\mathbf{k}_a|^{1/2} + |\mathbf{k}_b|^{1/2} = |\mathbf{k}_c|^{1/2} + |\mathbf{k}_d|^{1/2}.$$

In the (k_a, l_a)-plane, we can plot the level curves of $|\mathbf{k}_a|^{1/2} + |\mathbf{k}_b|^{1/2}$, with $k_b = 1 - k_a$ and $l_b = -l_a$; these are shown in Figure 4. Then, for $\mathbf{k}_a$ joining the origin of the plane to some point on a level curve, $\mathbf{k}_b$ joins this point to $(1, 0)$, $\mathbf{k}_c$ joins $(1, 0)$ to some other point on the same level curve, and $\mathbf{k}_d$ joins this point to the origin (Phillips, 1977).

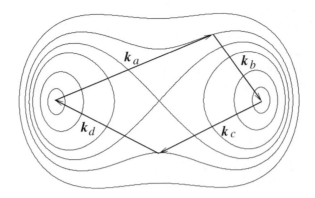

Figure 4. Resonant quartets for surface gravity waves.

6.1 Interaction equations

In contrast with wave triads, wave quartets are never completely isolated: this is because there are always trivial resonances: quartets made of a single wave, $(a, a, -a, -a)$, or of two waves $(a, -a, b, -b)$ obviously satisfy (6.2)–(6.3). Thus, the simplest truncation of the amplitude equations with cubic nonlinearity

$$\dot{A}_a = \frac{1}{6} \sum_{bcd} I_a^{bcd} A_b^* A_c^* A_d^* \exp(2i\Omega_{abcd}t)\delta_{\mathbf{k}_a+\mathbf{k}_b+\mathbf{k}_c+\mathbf{k}_d}$$

will be of the form

$$\dot{A}_a = -iA_a \sum_{j=a,d} J_a^j |A_j|^2 + I_a^{bcd} A_b^* A_c^* A_d^* \exp(2i\Omega_{abcd}t),$$

$$\dot{A}_b = -iA_b \sum_{j=a,d} J_b^j |A_j|^2 + I_b^{cda} A_c^* A_d^* A_a^* \exp(2i\Omega_{abcd}t),$$

$$\dot{A}_c = -iA_c \sum_{j=a,d} J_c^j |A_j|^2 + I_c^{dab} A_d^* A_a^* A_b^* \exp(2i\Omega_{abcd}t),$$

$$\dot{A}_d = -iA_d \sum_{j=a,d} J_d^j |A_j|^2 + I_d^{abc} A_a^* A_b^* A_c^* \exp(2i\Omega_{abcd}t).$$

Taking permutations of (a, b, c, d) into account gives

$$J_a^j = iI_a^{(-a)j(-j)} \quad \text{for} \quad j \neq a \quad \text{and} \quad J_a^a = iI_a^{(-a)a(-a)}/2. \tag{6.4}$$

Pseudoenergy conservation imposes that interactions coefficients J_a^j be real: the effect of the corresponding terms is therefore a nonlinear frequency shift, familiar from the study

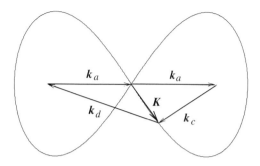

Figure 5. Wavevector configuration for surface-wave instability.

of single-wave propagation. Indeed, if a is the only mode excited, then its amplitude evolves according to

$$A_a(t) = A_{a0}\exp(-\mathrm{i}J_a^a|A_{a0}|^2 t), \tag{6.5}$$

with A_{a0} as initial value, corresponding to a wave propagating with the nonlinear frequency $\omega_a + J_a^a|A_{a0}|^2$.

Just as for triad interactions, the conservation properties of the original system impose constraints on the interaction coefficients I_a^{bcd}. For resonant quartets, the relation

$$\frac{E_a I_a^{bcd}}{\omega_a} = \frac{E_b I_b^{cda}}{\omega_b} = \frac{E_c I_c^{dab}}{\omega_c} = \frac{E_d I_d^{abc}}{\omega_d}$$

holds for sufficiently symmetric systems such as the one governing surface gravity waves (e.g. Simmons, 1969). Also, by scaling, the interaction coefficients can be reduced to the signature of the wave action E/ω. Like the triad equations, the quartet interaction equations can in fact be integrated in closed form in terms of elliptic functions.

6.2   Wave instability

In the absence of resonant triads, the instability of small-amplitude waves is controlled by resonant quartets. To examine the stability of a wave a in this case the pump-wave approximation can be used; this assumes that the amplitude of wave a remains given by (6.5) for all time. The quartet involves only three waves: the primary wave a, $b = a$, and the secondary waves c and d, with

$$\mathbf{k}_c = -\mathbf{k}_a + \mathbf{K} \quad \text{and} \quad \mathbf{k}_d = -\mathbf{k}_a - \mathbf{K}.$$

with $\mathbf{K}$ satisfying

$$2\Omega = 2\Omega(\mathbf{k}_a) - \omega(\mathbf{k}_a + \mathbf{K}) - \omega(\mathbf{k}_a - \mathbf{K}) \approx 0.$$

The wavevectors in this case are then as depicted in Figure 5.

Assuming that $A_{a0} = A_a(0)$ is real and taking permutations into account the two secondary waves are seen to obey the evolution equations

$$\dot{A}_c = -iJ_c^a A_c A_{a0}^2 + \frac{1}{2} I_c^{aad} A_d^* A_{a0}^2 \exp[2i(\Omega + J_a^a A_{a0}^2)t]$$

$$\dot{A}_d = -iJ_d^a A_d A_{a0}^2 + \frac{1}{2} I_d^{aac} A_c^* A_{a0}^2 \exp[2i(\Omega + J_a^a A_{a0}^2)t].$$

Solutions can be sought in the form

$$A_c = \hat{A}_c \exp[i(\Omega + J_a^a A_{a0}^2)t] \exp(\lambda t)$$

$$A_d = \hat{A}_d \exp[i(\Omega + J_a^a A_{a0}^2)t] \exp(\lambda^* t)$$

for constant $\hat{A}_c$, $\hat{A}_d$ and growth rate λ. The growth rate is found to be

$$\lambda = \frac{i(J_a^c - J_a^b)}{2} \pm \frac{1}{2} \left[I_c^{aad}(I_d^{aac})^* A_{a0}^4 - [2\Omega + (2J_a^a - J_c^a - J_d^a)A_{a0}^2]^2 \right]^{1/2}. \qquad (6.6)$$

Clearly, it can have a non-zero real part, corresponding to an instability of wave a if $I_c^{aad}(I_d^{aac})^* > 0$ and $\Omega/A_{a0}^2 + (2J_a^a - J_c^a - J_d^a)$ is small enough (see Craik, 1985, and references therein for application to surface gravity waves).

The limit $|\mathbf{K}| \ll |\mathbf{k}_a|$ is of particular interest: this is the situation termed side-band instability, in which the perturbation of wave a consists of two waves with wavevectors close to $\mathbf{k}_a$ and thus represents a large-scale modulation of the amplitude of a. In this limit, all the interaction coefficients involved in (6.6) are related to $I_a^{(-a)a(-a)}$ or, equivalently to

$$J = J_a^a = iI_a^{(-a)a(-a)}.$$

Noting that the property $A_{-a} = (A_a)^*$ implies that $I_{-a}^{(-b)(-c)(-d)} = (I_a^{bcd})^*$, we find

$$I_c^{aad} \sim I_d^{aac} \sim 2iJ \quad \text{and} \quad J_c^a \sim J_c^d \sim -2J$$

as $|\mathbf{K}| \to 0$ and hence (6.6) reduces to

$$\lambda = \left[2\Omega J A_{a0}^2 - \Omega^2 \right]^{1/2},$$

with

$$\Omega = -\frac{1}{2}\mathbf{K} \cdot \frac{\partial^2 \omega}{\partial k \partial k}(\mathbf{k}_a) \cdot \mathbf{K}. \qquad (6.7)$$

Side-band instability occurs when

$$\Omega(2J A_{a0}^2 - \Omega) > 0. \qquad (6.8)$$

This is a simple criterion, which involves only the nonlinear frequency shift of wave a (through J), and the Hessian of the dispersion relation (see Whitham, 1965); it can also be obtained from the nonlinear Schrödinger equation modelling the evolution of large-scale modulations of a wave (cf. Chapter 2). It is satisfied for surface gravity waves which are observed to be unstable to side bands (the mechanism is then termed Benjamin–Feir instability) as well as to perturbations with finite $\mathbf{K}$.

Klein–Gordon equation. The nonlinear Klein–Gordon equation

$$\psi_{tt} - \psi_{xx} + \psi + 4\sigma\psi^3 = 0, \quad \sigma = \pm 1. \tag{6.9}$$

provides a simple application of the theory above (Whitham, 1974). Linear waves satisfy the dispersion relation

$$\omega = \pm(1 + k^2)^{1/2}. \tag{6.10}$$

The nonlinear interaction equations are readily derived by substituting the expansion $\psi = \sum_a A_a(t) \exp[i(kx - \omega t)]$ into (6.9). For $A_a = O(\epsilon)$, the evolution takes place over a time scale $t = O(\epsilon^2)$; the terms involving $\ddot{A}_a$ are $O(\epsilon^5)$ and can be neglected, leading to

$$\dot{A}_a = -\frac{2\sigma i}{\omega_a} \sum_{abc} A_b^* A_c^* A_d^* \exp(2i\Omega_{abcd}t), \tag{6.11}$$

that is, to interaction equations of the general form with interaction coefficients $I_a^{bcd} = -12\sigma i/\omega_a$. From (6.4), the coefficient $J = J_a$ governing the frequency shift of wave a is $J = 6\sigma/\omega_a$. From (6.7), we compute $\Omega = -K^2/(2\omega_a^3)$. The condition (6.8) for side-band instability is therefore $12\sigma A_{a0}^2 + K^2/2\omega_a^2 < 0$; that is, side-band instability requires $\sigma < 0$ and involves only secondary waves satisfying

$$K^2 < 24|\sigma|\omega_a^2 A_{a0}^2.$$

7 Alternative derivations

The derivation of the interaction equations described in section 3 uses the conservation properties of the system (2.7) only through the orthogonality relations (2.11). As a result, the symmetry properties of the interaction coefficients (3.9)–(3.10) are not explicit: the conservation laws need to be invoked to establish them, although they can be confirmed by (tedious) manipulations. There are alternative approaches to the derivation of the interaction equations which rely explicitly on the Lagrangian or Hamiltonian structure which underlies the conservative properties. We describe such approaches in this section.

7.1 Lagrangian systems

For systems that can be derived from a variational principle,

$$\delta \int dt \int d\mathbf{x}\, L(\mathbf{u}, \dot{\mathbf{u}}, \nabla\mathbf{u}, \cdots) = 0,$$

the interaction equations are conveniently derived by introducing the expansion (2.10) into the Lagrangian density $L(\mathbf{u}, \dot{\mathbf{u}}, \nabla\mathbf{u}, \cdots)$, performing the integration over space, and taking the variations of the resulting functional with respect to the amplitudes. Typically, one obtains a variational principle of the form

$$\delta \int \left(\mathcal{L}^{(2)} + \mathcal{L}^{(3)} + \cdots \right) dt = 0,$$

where

$$\mathcal{L}^{(2)} = \sum_a D(\mathbf{k}_a, \omega_a)|A_a|^2,$$

for some function $D(k, \omega)$, and

$$\mathcal{L}^{(3)} = \mathrm{i} \int \mathrm{d}t \left[\sum_a D_\omega(\mathbf{k}_a, \omega_a) \dot{A}_a A_a^* - \frac{1}{3} \sum_{abc} \Gamma_{abc}^* A_a A_b A_c \exp(-2\mathrm{i}\Omega_{abc}t)\delta_{\mathbf{k}_a + \mathbf{k}_b + \mathbf{k}_c} \right],$$

with Γ_{abc} constants symmetric in (a, b, c). The variations with respect to the amplitudes A_a give at leading order

$$D(\mathbf{k}_a, \omega_a) = 0.$$

The function $D(k, \omega)$ is therefore a form of the dispersion relation. At next order, we find

$$D_\omega(\mathbf{k}_a, \omega_a)\dot{A}_a = \frac{1}{2} \sum_{bc} \Gamma_{abc} A_b^* A_c^* \exp(2\mathrm{i}\Omega_{abc}t)\delta_{\mathbf{k}_a + \mathbf{k}_b + \mathbf{k}_c}.$$

This is a result identical to (3.6), with the interaction coefficients

$$I_a^{bc} = \frac{\Gamma_{abc}}{D_\omega(\mathbf{k}_a, \omega_a)} \tag{7.1}$$

displaying a symmetric form. It can be shown that the pseudoenergy of a mode is given by

$$E_a = \omega_a D_\omega(\mathbf{k}_a, \omega_a)$$

for a suitable normalization of the Lagrangian (cf. Craik, 1985). The properties (3.10) are thus direct consequences of the form (7.1) of the interaction coefficients; in this variational formulation, they hold for non-resonant as well as resonant interactions.

Klein–Gordon equation. The Klein–Gordon equation (6.9) derives from the variational principle

$$\delta \iint \left[\psi_t^2 - \psi_x^2 - \psi^2 - 2\sigma\psi^4 \right] \mathrm{d}x \, \mathrm{d}t = 0.$$

This can be used to re-derive the interaction equations (6.11). Introducing the expansion $\psi = \sum_a A_a(t) \exp[\mathrm{i}(kx - \omega t)]$ gives the variational principle

$$\delta \int \left(\mathcal{L}^{(2)} + \mathcal{L}^{(4)} + \cdots \right) \mathrm{d}t,$$

where

$$\mathcal{L}^{(2)} = (2\pi)^2 \sum_a (\omega^2 - k_a^2 - 1)|A_a|^2$$

and

$$\mathcal{L}^{(4)} = (2\pi)^2 \left[\sum_a 2\mathrm{i}\omega_a \dot{A}_a A_a^* - 2\sigma \sum_{abcd} A_a^* A_b^* A_c^* A_d^* \exp(2\mathrm{i}\Omega_{abcd}t) \right].$$

Taking the variation of $\mathcal{L}^{(2)}$ with respect to A_a gives the dispersion relation relation (6.10). Taking the variation of $\mathcal{L}^{(3)}$ yields (6.11).

The Lagrangian formulation provides the most convenient route for the derivation of interaction equations for surface (capillary–)gravity wave. For Rossby waves and internal gravity waves, it is somewhat awkward to apply, because the relevant equations of motion in their usual form (such as (2.1)) cannot be derived from a variational principle. Variational formulations exist, but they require the use of somewhat esoteric variables (Lagrangian labels (Salmon, 1988; Holm and Zeitlin, 1998) or Clebsch variables (Seliger and Whitham, 1968)). An alternative is to use the Hamiltonian formulation of the equation of motion; some preliminary work is however needed to recast the equations of motion as a near-canonical systems. We now briefly describe how this can be done for Rossby waves as described by (2.1), following Zakharov and Piterbarg (1988) (Zakharov et al., 1992; Zeitlin, 1992, see also).

7.2 Hamiltonian formulation for Rossby waves

The idea of Zakharov and Piterbarg (1988) is to introduce a near-identify transformation of the dependent variable, here the vorticity ζ, that gives a simple form to the equation of motion. The change of variables can in principle be carried out to arbitrary high order of accuracy; since we are interested in deriving only the quadratic terms of the interaction equations, we systematically neglect the $O(\epsilon^3)$ terms.

Consider the new dependent variable $\tilde{\zeta}$ defined by

$$\tilde{\zeta} = \zeta + \zeta\zeta_y/\beta. \tag{7.2}$$

For small ζ (in fact for $|\zeta_y/\beta| \ll 1$), this transformation can be inverted according to

$$\zeta = \tilde{\zeta} - \tilde{\zeta}\tilde{\zeta}_y/\beta + O(\zeta^3). \tag{7.3}$$

The evolution equation (2.1) for ζ is easily seen to imply that $\tilde{\zeta}$ obeys

$$\tilde{\zeta}_t + \beta\partial_x(\psi + \psi_y\tilde{\zeta}) = O(\epsilon^3).$$

This has the Hamiltonian form

$$\tilde{\zeta}_t = \beta\partial_x\frac{\partial\mathcal{H}}{\delta\tilde{\zeta}}, \quad \text{with} \quad \mathcal{H} = \frac{1}{2}\int |\nabla\psi|^2\,\mathrm{dx} = -\frac{1}{2}\int \psi\zeta\,\mathrm{dx}.$$

From this form amplitude equations are readily derived: let

$$\tilde{\zeta} = \sum_a C_a(t)\exp(i\mathbf{k}_a \cdot \mathbf{x}) \quad \text{with} \quad C_a = \frac{1}{(2\pi)^2}\int \tilde{\zeta}(\mathbf{x},t)\exp(-i\mathbf{k}_a \cdot \mathbf{x})\,\mathrm{dx}. \tag{7.4}$$

By differentiation, we find the evolution equations for the amplitudes C_a to be

$$\dot{C}_a = \frac{\beta}{(2\pi)^2}\int \partial_x\left(\frac{\delta\mathcal{H}}{\delta\zeta}\right)\exp(-i\mathbf{k}_a \cdot \mathbf{x})\,\mathrm{dx}.$$

Noting from (7.4) that

$$\frac{\delta \mathcal{H}}{\delta \zeta} = \frac{1}{(2\pi)^2} \sum_a \frac{\partial \mathcal{H}}{\partial C_a} \exp(-i\mathbf{k}_a \cdot \mathbf{x}),$$

these reduce to the Hamiltonian equations

$$\dot{C}_a = \frac{i\beta k_a}{(2\pi)^2} \frac{\partial \mathcal{H}}{\partial C_a^*}. \tag{7.5}$$

The vorticity and streamfunction are

$$\zeta = \sum_a C_a(t) \exp(i\mathbf{k}_a \cdot \mathbf{x}) - \frac{i}{2\beta} \sum_{bc} (l_b + l_c) C_b C_c \exp[i(\mathbf{k}_b + \mathbf{k}_c) \cdot \mathbf{x}],$$

$$\psi = \sum_a \frac{-1}{k_a^2 + l_a^2} C_a(t) \exp(i\mathbf{k}_a \cdot \mathbf{x})$$

$$+ \frac{i}{2\beta} \sum_{bc} \frac{l_b + l_c}{(k_b + k_c)^2 + (l_b + l_c)^2} C_b C_c \exp[i(\mathbf{k}_b + \mathbf{k}_c) \cdot \mathbf{x}],$$

with the Hamiltonian expressed in terms of the C_a as

$$\mathcal{H} = \frac{(2\pi)^2}{2} \left(\sum_a \frac{|C_a|^2}{k_a^2 + l_a^2} + \frac{i}{3\beta} \sum_{abc} \Delta C_a^* C_b^* C_c^* \right),$$

where, after symmetrization,

$$\Delta_{abc} = -\left(\frac{l_a}{k_a^2 + l_a^2} + \frac{l_b}{k_b^2 + l_b^2} + \frac{l_c}{k_c^2 + l_c^2} \right).$$

With these results, the interaction equations are found to be

$$\dot{C}_a = -i\omega_a C_a - k_a \sum_{bc} \Delta_{abc} C_b^* C_c^*.$$

With $C_a = A_a \exp(-i\omega_a t)$, these are equivalent to (3.2) near resonance since

$$-k_a \Delta_{abc} \approx \frac{k_a l_b}{k_b^2 + l_b^2} + \frac{k_a l_c}{k_c^2 + l_b^2} - \frac{l_a k_b}{k_b^2 + l_b^2} - \frac{l_a k_c}{k_c^2 + l_c^2} = I_a^{bc}$$

where I_a^{bc} is given in (3.3).

8 Concluding remarks

In these lectures, we have assumed that the equations of motion (2.7) have no explicit dependence on the coordinates. As a result, the waves have a simple spatial structure, proportional to $\exp(i\mathbf{k} \cdot \mathbf{x})$. This is a major restriction, of course, which it is often necessary to relax: explicit dependence on some of the coordinates appear for instance when the domain geometry is not Cartesian, or for waves propagating in sheared fluid flows.

When there is an explicit coordinate dependence, the construction of sections 2–3 can formally be carried out with only minor modifications. For instance, if the second coordinate y appears explicitly in a two dimensional system, modal solutions of the form

$$\mathbf{u} = A\hat{\mathbf{u}}(y)\exp[\mathrm{i}(kx - \omega t)]$$

can be sought. The dispersion relation and y-dependent eigenvector are found by solving a differential (rather than matrix) eigenvalue problem; and relations such as (2.11) continue to hold, but with an additional integration over y on the right-hand side. An important change is that the interaction condition (3.4) has a reduced dimension, since it involves scalar wavenumbers instead of two-dimensional wavevectors. For domains finite in the y direction, one can expect an infinite discrete set of frequencies and eigenvectors $\hat{\mathbf{u}}$ for each k, that is, the dispersion relation has an infinite number of branches. (The label a in the expansion (3.5) and b, c in (3.6) then involves both the wavenumber k_a and the branch index.) For examples of computation of interaction coefficients in such a situation see, for instance, Reznik et al. (1993) (Rossby waves on a sphere), or Vanneste and Vial (1994) (Rossby waves and internal gravity waves in shear flows).

For shear flows, this is not the whole story, however. In addition to discrete frequencies, the eigenvalue problem for ω and $\hat{\mathbf{u}}(y)$ also admits a continuous spectrum for all phase speed ω/k in the range of the flow velocity, with corresponding $\hat{\mathbf{u}}(y)$ singular at the so-called critical level, where the phase speed ω/k matches the flow velocity (e.g., Case, 1960; Balmforth and Morrison, 2002). Little is known about the nonlinear interactions involving these singular modes. These are potentially important, in particular because they could lead to nonlinear instabilities through explosive interactions (Becker and Grimshaw, 1993).

Acknowledgements. It is a pleasure to thank R. Grimshaw for organizing the enjoyable summer school for which these lecture notes were prepared.

Bibliography

M. S. Alber, G. L. Luther, J. E. Marsden, and J. M. Robbins. Geometric phase, reduction and Lie–Poisson structure for the resonant three-wave interactions. *Physica D*, 123: 271–290, 1998.

N. J. Balmforth and P. J. Morrison. Hamiltonian description of shear flow. In *Large-scale atmosphere-ocean dynamics, Vol. II*, pages 117–142. Cambridge Univ. Press, 2002.

J. M. Becker and R. Grimshaw. Explosive resonant triads in a continuously stratified shear flows. *J. Fluid Mech.*, 257:219–228, 1993.

F. P. Bretherton. Resonant interactions between waves. The case of discrete oscillations. *J. Fluid Mech.*, 20:457–479, 1964.

K. M. Case. Stability of inviscid Couette flow. *Phys. Fluids*, 3:143–148, 1960.

A. D. D. Craik. *Wave interactions and fluid flows*. Cambridge University Press, 1985.

A. D. D. Craik and J. A. Adam. Explosive resonant interaction in a three-layer fluid flow. *J. Fluid Mech.*, 92:15–33, 1979.

A. E. Gill. *Atmosphere-ocean dynamics.* Academic Press, 1982.

R. Grimshaw. Wave action and wave–mean flow interaction, with application to stratified shear flows. *Ann. Rev. Fluid Mech.*, 16:11–44, 1984.

K. Hasselmann. A criterion for nonlinear wave stability. *J. Fluid Mech.*, 30:737–739, 1967.

D. D. Holm and V. Zeitlin. Hamilton's principle for quasigeostrophic motion. *Phys. Fluids*, 10:800–806, 1998.

S. Ichimaru. *Basic principles of plasma physics: a statistical approach.* W.A. Benjamin, 1973.

E. L. Ince. *Ordinary differential equations.* Dover, 1956.

D. J. Kaup, A. Reiman, and A. Bers. Space-time evolution of nonlinear three-wave interactions. I. interactions in a homogeneous medium. *Rev. Mod. Phys.*, 51:275–309, 1979. Erratum: pp. 915–917.

P. J. Morrison. Hamiltonian description of the ideal fluid. *Rev. Mod. Phys.*, 70:467–521, 1998.

J. Pedlosky. *Geophysical fluid dynamics.* Springer–Verlag, 1987.

O. M. Phillips. *The dynamics of the upper ocean.* Cambridge University Press, 2nd edition, 1977.

G. M. Reznik, L. I. Piterbarg, and E. A. Kartashova. Nonlinear interactions of spherical Rossby modes. *Dyn. Atmos. Oceans*, 18:235–252, 1993.

P. Ripa. On the theory of nonlinear wave-wave interactions among geophysical waves. *J. Fluid Mech.*, 103:87–115, 1981.

R. Salmon. Hamiltonian fluid mechanics. *Ann. Rev. Fluid Mech.*, 20:225–256, 1988.

R. L. Seliger and G. B. Whitham. Variational principle in continuous mechanics. *Proc. R. Soc. Lond. A*, 305:1–25, 1968.

T. G. Shepherd. Symmetries, conservation laws and Hamiltonian structure in geophysical fluid dynamics. *Adv. Geophys.*, 32:287–338, 1990.

W.F. Simmons. A variational method for weak resonant wave interactions. *Proc. R. Soc. Lond. A*, 309:551–575, 1969.

J. Vanneste. Explosive resonant interaction of baroclinic Rossby waves and stability of multilayer quasi-geostrophic flow. *J. Fluid Mech.*, 291:83–107, 1995.

J. Vanneste and T. G. Shepherd. On wave action and phase in the non-canonical Hamiltonian formulation. *Proc. R. Soc. Lond. A*, 455:3–21, 1999.

J. Vanneste and F. Vial. On the nonlinear interaction of geophysical waves in shear flows. *Geophys. Astrophys. Fluid Dynam.*, 78:115–141, 1994.

G. B. Whitham. Nonlinear dispersive waves. *Proc. R. Soc. Lond. A*, 283:238–261, 1965.

G. B. Whitham. *Linear and nonlinear waves.* Wiley, 1974.

V. E. Zakharov, V.S. L'vov, and G. Falkovich. *Kolmogorov spectra of turbulence I: wave turbulence.* Springer–Verlag, 1992.

V.E. Zakharov and L.I. Piterbarg. Canonical variables for Rossby waves and plasma drift waves. *Phys. Lett. A*, 126:497–500, 1988.

V. Zeitlin. Vorticity and waves: geometry of phase-space and the problem of normal variables. *Phys. Lett. A*, 164:177–183, 1992.

Wave–mean interaction theory

Oliver Bühler

Courant Institute of Mathematical Sciences
New York University, New York, NY 10012, U.S.A.
obuhler@cims.nyu.edu

Abstract.
This is an informal account of the fluid-dynamical theory describing nonlinear interactions between small-amplitude waves and mean flows. This kind of theory receives little attention in mainstream fluid dynamics, but it has been developed greatly in atmosphere and ocean fluid dynamics. This is because of the pressing need in numerical atmosphere–ocean models to approximate the effects of unresolved small-scale waves acting on the resolved large-scale flow, which can have very important dynamical implications. Several atmosphere ocean example are discussed in these notes (in particular, see §5), but generic wave–mean interaction theory should be useful in other areas of fluid dynamics as well.

We will look at a number of examples relating to the basic problem of classical wave–mean interaction theory: finding the nonlinear $O(a^2)$ mean-flow response to $O(a)$ waves with *small amplitude* $a \ll 1$ in *simple geometry*. Small wave amplitude $a \ll 1$ means that the use of linear theory for $O(a)$ waves propagating on an $O(1)$ background flow is allowed. Simple geometry means that the flow is periodic in one spatial coordinate and that the $O(1)$ background flow does not depend on this coordinate. This allows the use of averaging over the periodic coordinate, which greatly simplifies the problem.

1 Two-dimensional incompressible homogeneous flow

This is our basic starting point. We first develop the mathematical equations for this kind of flow and then we consider waves and mean flows in it.

1.1 Mathematical equations

We work in a flat, two-dimensional domain with Cartesian coordinates $\boldsymbol{x} = (x, y)$ and velocity field $\boldsymbol{u} = (u, v)$. In the y-direction the domain is bounded at $y = 0$ and $y = D$ by solid impermeable walls such that $v = 0$ there. In the x-direction there are periodic boundary conditions such that $\boldsymbol{u}(x + L, y, t) = \boldsymbol{u}(x, y, t)$. In an atmospheric context we can think of x as the "zonal" (i.e. east–west) coordinate and of y as the "meridional" (i.e. south–north) coordinate. The period length L is then the Earth's circumference.

The flow is incompressible, which means that the velocity field is area-preserving and hence has zero divergence:

$$\boldsymbol{\nabla} \cdot \boldsymbol{u} = 0 \quad \Leftrightarrow \quad u_x + v_y = 0. \tag{1.1}$$

The velocity field induces a time derivative following the fluid flow, which is called the *material* derivative

$$\frac{D}{Dt} \equiv \frac{\partial}{\partial t} + u\frac{\partial}{\partial x} + v\frac{\partial}{\partial y} = \frac{\partial}{\partial t} + (\boldsymbol{u} \cdot \boldsymbol{\nabla}) \quad . \tag{1.2}$$

Evaluating the material derivative of any flow variable at location $\boldsymbol{x}$ and time t gives the rate of change of this variable as experienced by the fluid particle that is at location $\boldsymbol{x}$ at the time t. The quadratic nonlinearity of the material derivative when applied to $\boldsymbol{u}$ itself gives fluid dynamics its peculiar mathematical flavour.

The momentum equations for inviscid ideal flow is provided by Newton's law as

$$\frac{D\boldsymbol{u}}{Dt} + \frac{1}{\rho}\boldsymbol{\nabla}p = 0, \tag{1.3}$$

where ρ is the fluid density per unit area and p is the pressure. We assume that the flow is homogeneous (i.e. $\boldsymbol{\nabla}\rho = 0$) and hence ρ can be absorbed in the definition of p so that we can set $\rho = 1$ throughout. The mathematical problem is completed by specifying the boundary conditions mentioned before:

$$\boldsymbol{u}(x + L, y, t) = \boldsymbol{u}(x, y, t), \qquad v = 0, \, p_y = 0 \tag{1.4}$$
$$p(x + L, y, t) = p(x, y, t) \qquad \text{at } y = \text{ and } y = D. \tag{1.5}$$

The boundary condition for the pressure at the wall follows from evaluating the y-component of (1.3) at the wall, where $v = 0$:

$$v_t + uv_x + vv_y + p_y = 0 \quad \Rightarrow \quad p_y = 0. \tag{1.6}$$

Together with (1.1) we now have three equations for the three variables u, v, p.

However, the pressure is not really an independent variable. This is a peculiarity of incompressible flow and can be seen as follows. Taking the divergence of (1.3) results in

$$\nabla^2 p = -\boldsymbol{\nabla} \cdot \boldsymbol{u}_t - \boldsymbol{\nabla} \cdot [(\boldsymbol{u} \cdot \boldsymbol{\nabla})\boldsymbol{u}] = -\boldsymbol{\nabla} \cdot [(\boldsymbol{u} \cdot \boldsymbol{\nabla})\boldsymbol{u}] \tag{1.7}$$

due to (1.1). Hence (1.7) is a Poisson equation for p in terms of the velocities, which can be solved for p. This means that $\boldsymbol{u}$ *determines* the pressure p instantaneously at any given moment in time. In other words, we can only specify initial conditions for $\boldsymbol{u}$ but not for p.

It turns out that we can eliminate p at the outset, which reduces the number of variables that need to be considered. To do this we take the curl of (1.3), which eliminates the pressure gradient term. This brings in the *vorticity* vector $\boldsymbol{\nabla} \times \boldsymbol{u}$, which is

$$\boldsymbol{\nabla} \times \boldsymbol{u} = (0, 0, v_x - u_y) \tag{1.8}$$

in two dimensions. There is only one nonzero component, which we will denote by

$$q \equiv v_x - u_y. \tag{1.9}$$

So, subtracting the y-derivative of the x-component of (1.3) from the x-derivative of its y-component leads to

$$\frac{Dq}{Dt} + q\boldsymbol{\nabla}\cdot\boldsymbol{u} + p_{yx} - p_{xy} = 0 \quad \Rightarrow \quad \frac{Dq}{Dt} = 0. \tag{1.10}$$

This means that q is advected by the flow. We say that q is a *material invariant*.

We can satisfy (1.1) exactly by introducing a *stream function* ψ such that

$$u = -\psi_y, \quad v = +\psi_x. \tag{1.11}$$

Clearly, ψ is determined only up to an arbitrary constant. The relationship between q and ψ is

$$q = v_x - u_y = \psi_{xx} + \psi_{yy} = \nabla^2\psi. \tag{1.12}$$

For given q this equation can be inverted to find ψ, though some care is needed because our channel domain is doubly connected. This means we require boundary conditions on ψ at both walls (in addition to requiring ψ to be x-periodic with period L). At the walls $\psi_x = 0$ and hence ψ is a constant there. We can set $\psi = 0$ at $y = 0$ without loss of generality due to the arbitrary constant in ψ. Then we have

$$\psi|_{y=D} = -\int_0^D u\,dy = A(t). \tag{1.13}$$

The evolution of $A(t)$ has to be determined from (1.3). This gives $A =$const. and so A is determined once and for all from the initial conditions. Physically, $A \neq 0$ corresponds to a uniform, vorticity-free flow along the channel. Together, q and A determine ψ uniquely in our channel domain. With this understood, we will not consider A explicitly from now on.

So, in summary, if we take q, ψ as our basic two variables then we have the closed system of two equations

$$\nabla^2\psi = q \tag{1.14}$$

and

$$\frac{Dq}{Dt} = 0, \quad \Leftrightarrow \quad q_t + uq_x + vq_y = 0, \quad \Leftrightarrow \quad q_t - \psi_y q_x + \psi_x q_y = 0. \tag{1.15}$$

This is called the *vorticity–stream function* formulation of two-dimensional fluid dynamics. The material invariance of q allows many analytical simplifications, as we will see. Initial conditions can be specified either in q or in ψ; one can compute one from the other via (1.14).

1.2   Waves on shear flows

What is the linear dynamics of the system (1.14-1.15) relative to a state of rest? The first equation is already linear and the linear part of the second is simply

$$q_t = 0. \tag{1.16}$$

This means that *any* vorticity distribution is a steady solution of the linearized equations. So the linear state is infinitely degenerate and has no dynamics: all dynamics is necessarily nonlinear in the present case.[1] However, this changes when we consider the linear dynamics relative to a shear flow along the channel. Specifically, we consider an $O(1)$ shear flow

$$\boldsymbol{U}(y) = (U(y), 0) \tag{1.17}$$

with vorticity

$$Q(y) = -U_y. \tag{1.18}$$

It is easy to check that this gives a trivial steady state for all profiles $U(y)$.

We now consider linear waves on top of this shear flow. The wave amplitude is denoted by a suitable non-dimensional positive number $a \ll 1$ such that

$$\begin{aligned}
\boldsymbol{u} &= \boldsymbol{U} + \boldsymbol{u}' + O(a^2) & (1.19)\\
q &= Q + q' + O(a^2) & (1.20)\\
\psi &= \Psi + \psi' + O(a^2) & (1.21)
\end{aligned}$$

where $\{\boldsymbol{u}', q', \psi'\}$ are all understood to be $O(a)$ and Ψ is the stream function belonging to $\boldsymbol{U}$. So we have expanded the flow into an $O(1)$ background flow, $O(a)$ waves, and as yet unspecified further $O(a^2)$ terms. It is straightforward to show that (1.14-1.15) yield

$$\nabla^2 \psi' = q' \quad \text{and} \tag{1.22}$$

$$\left(\frac{\partial}{\partial t} + U \frac{\partial}{\partial x} \right) q' + v' Q_y = 0 \tag{1.23}$$

at $O(a)$. The operator acting on q' in the second equation gives the time derivative along $O(1)$ material trajectories. We will use the short-hand

$$\mathrm{D}_t \equiv \left(\frac{\partial}{\partial t} + U \frac{\partial}{\partial x} \right) \tag{1.24}$$

for it. The equation itself expresses that q' along these trajectories changes due to advection of particles in the y-direction in the presence of an $O(1)$ vorticity gradient Q_y.

The system (1.22-1.23) (and its viscous counterpart) has been studied for a long time (e.g. Drazin and Reid (1981)), mainly in order to find unstable growing modes, i.e. complex-valued modes of the form $\psi' \propto \exp(i(kx - ct))\hat{\psi}(y)$ with a nonzero imaginary part of the phase speed c. The modal approach has many practical advantages, but clarity and physical insight are not among them. We will take a different approach to this system, which lends itself to far-reaching generalizations. As a bonus, we will derive Rayleigh's theorem (one of the main results of modal theory) *en passant*.

First, we introduce the helpful linear particle displacement η' in the y-direction via

$$\mathrm{D}_t \eta' = v'. \tag{1.25}$$

---

[1] This peculiar fact gives rise to the popular quip that even linear fluid dynamics is more complicated than quantum mechanics! This always raises a laugh, especially among fluid dynamicists.

So the rate of change of η' along $O(1)$ material trajectories is given by v'. Of course, a complete specification of η' requires initial conditions as well. Now, combining (1.23) and (1.25) gives

$$D_t\{q' + \eta' Q_y\} = 0, \qquad (1.26)$$

where we have used that $D_t Q_y = 0$. This means that if η' is initialized to be

$$\eta' = -\frac{q'}{Q_y} \quad \text{at } t = 0 \qquad (1.27)$$

then this relation will hold at all later times as well.

We now introduce the important concept of *zonal averaging* along the channel: for any field $A(x, y, t)$ we define

$$\overline{A} \equiv \frac{1}{L} \int_{-L/2}^{+L/2} A(x+s, y, t)\,ds \qquad (1.28)$$

to be the *mean* part of A. For x-periodic A, the mean $\overline{A}$ is simply its x-average at fixed y and t, and then $\overline{A}$ does not depend on x. However, this particular definition of $\overline{A}$ has the advantage that $\overline{(x, y, t)} = (x, y, t)$, which is a useful property. Averaging is a linear operation and hence

$$\overline{A + B} = \overline{A} + \overline{B} \qquad (1.29)$$

holds for all A, B. Furthermore, averaging commutes with taking partial derivatives in space and time (it also commutes with D_t though it does not in general commute with the full material derivative). Most importantly, this implies that

$$\overline{A_x} = (\overline{A})_x = 0 \qquad (1.30)$$

for *all* x-periodic functions A. Clearly, averaging introduces an x-symmetry in $\overline{A}$ that need not have been present in A. We can now define the *disturbance* part of A to be

$$A' \equiv A - \overline{A} \quad \text{such that} \quad \overline{A'} = 0. \qquad (1.31)$$

This is the *exact* definition of the disturbance A', i.e. this definition holds without restriction to small-amplitude disturbances.

Combining (1.28) and (1.31) we note that averaging nonlinear terms results in a mixture of mean and disturbance parts. Specifically, the mean of a quadratic term is easily shown to be

$$\overline{AB} = \overline{A}\,\overline{B} + \overline{A'B'} \qquad (1.32)$$

after using $\overline{A'\overline{B}} = \overline{A'}\,\overline{B} = 0$ etc. This is the most important nonlinear average in fluid dynamics and it is exact, i.e. not restricted to small wave amplitudes.

Let us pause for a second to consider the two distinct mathematical tools we are using: zonal averaging and small-amplitude expansions. For instance, consider the explicit vorticity small-amplitude expansion

$$q = Q + q_1 + q_2 + \ldots + q_n + O(a^{n+1}), \qquad (1.33)$$

where the expansion subscripts mean that $q_n = O(a^n)$ (except for the $O(1)$ background term). Each of these terms can be decomposed into a mean and a disturbance part relative to the averaging operation:

$$q_n = \overline{q_n} + q_n'. \tag{1.34}$$

Now, the background vorticity has no disturbance part (i.e. $Q' = 0$, or $Q = \overline{Q}$) whilst the linear, first-order vorticity has no mean part: $\overline{q_1} = 0$, or $q_1 = q_1'$. Starting with q_2 all terms usually have both mean and disturbance parts. We can see now that strictly speaking q' and the other disturbance variables in (1.19) should be denoted by q_1' etc. However, we will usually suppress the cumbersome expansion subscripts when the meaning is clear from the context. From time to time we will use the expansion subscripts to highlight the nature of a particular approximation.

Returning to (1.23) now, we perform a crucial operation: multiply (1.23) by η' and then take the zonal average of that equation. This yields

$$\overline{\eta' D_t q'} + \overline{\eta' v'} Q_y = 0 \quad \Rightarrow \quad D_t \left(\frac{\overline{\eta' q'}}{2} \right) - \overline{q' v'} = 0 \tag{1.35}$$

after using (1.27). The first term is the time derivative of a new important variable, the zonal **pseudomomentum** per unit mass

$$\boxed{ \mathsf{p}(y,t) \equiv \frac{\overline{\eta' q'}}{2} = -\frac{\overline{\eta'^2}}{2} Q_y = -\frac{\overline{q'^2}}{2 Q_y} }. \tag{1.36}$$

The dimensions of p are that of a velocity and the second equality in (1.36) is useful because it makes clear that the sign of p is opposite to that of Q_y. The second term in (1.35) can be manipulated as follows:

$$-\overline{q'v'} = -\overline{(v_x' - u_y')v'} = -\overline{v_x' v'} + \overline{u_y' v'} \tag{1.37}$$

$$= -\frac{1}{2}\overline{(v'^2)}_x + \overline{(u'v')}_y - \overline{u'v_y'} \tag{1.38}$$

$$= 0 + \overline{(u'v')}_y + \overline{u'u_x'} = \overline{(u'v')}_y. \tag{1.39}$$

This made use of the continuity equation $u_x' + v_y' = 0$, integration by parts, and the key symmetry (1.30). Noting that $D_t \mathsf{p} = \mathsf{p}_t$ finally gives the pseudomomentum evolution equation

$$(\mathsf{p}_2)_t + (\overline{u_1' v_1'})_y = 0. \tag{1.40}$$

The pseudomomentum is an $O(a^2)$ quantity, but it is completely determined by the linear, $O(a)$ wave fields, i.e.

$$\mathsf{p}_2 = \frac{\overline{\eta_1' q_1'}}{2} \tag{1.41}$$

if expansion subscripts are used. Such $O(a^2)$ quantities are called *wave properties* to distinguish them from other $O(a^2)$ quantities (such as q_2) that depend on more than just the linear equations. Now, (1.40) yields a conservation law for the total, channel-integrated pseudomomentum, which is

$$\mathcal{P}(t) \equiv \int_0^L \int_0^D \mathsf{p}\, dx dy = L \int_0^D \mathsf{p}\, dy. \tag{1.42}$$

The time derivative of $\mathcal{P}$ is

$$\frac{d\mathcal{P}}{dt} = L \int_0^D \mathsf{p}_t\, dy = -L\, \overline{u'v'}\big|_{y=0}^{y=D} = 0, \tag{1.43}$$

because $v' = 0$ at the channel walls. Therefore the total pseudomomentum $\mathcal{P}$ is constant, which is a new conservation law.

This conservation law leads directly to Rayleigh's famous instability criterion, namely that the existence of an unstable normal mode for a given $U(y)$ implies that the vorticity gradient $Q_y = -U_{yy}$ must change sign somewhere in the domain. This follows from

$$\mathcal{P} = L \int_0^D -\frac{\overline{\eta'^2}}{2} Q_y\, dy = \text{const.} \tag{1.44}$$

and the fact that for a growing normal mode (which has constant shape in η' but grows in amplitude) the profile

$$\overline{\eta'^2}(y,t) = \exp(2\alpha t)\, \overline{\eta'^2}(y,0) \tag{1.45}$$

for some growth rate $\alpha > 0$. (For a normal mode α is proportional to the imaginary part of c.) This means that

$$\mathcal{P}(t) = \exp(2\alpha t)\, \mathcal{P}(0), \tag{1.46}$$

which is compatible with $\mathcal{P} = \text{const.}$ only if $\mathcal{P} = 0$. Therefore for a growing normal mode $\mathcal{P} = 0$ and (1.44) then implies that Q_y must change its sign somewhere in the domain. This is Rayleigh's famous theorem.

It can be noted that although a sign-definite Q_y therefore implies stability of normal modes, it does not preclude the localized transient growth of non-normal modes (e.g. Haynes (1987)).

1.3   Mean-flow response

We now consider the leading-order mean-flow response to the waves. This response occurs due to the quadratic nonlinearity of the equations, which produces a leading-order response at $O(a^2)$. However, as was first shown by Reynolds, in simple geometry it is often possible to write down a mean-flow equation that holds at finite amplitude. We will do this first and then specialize to $O(a^2)$. Substituting the exact decomposition $\boldsymbol{u} = \overline{\boldsymbol{u}} + \boldsymbol{u}'$ into the continuity equation and averaging yields

$$\overline{u_x} + \overline{v_y} = 0 \quad \Rightarrow \quad \overline{v} = \text{const.} = 0 \tag{1.47}$$

The last equality comes from the fact that at the walls $v = 0$. So we see that $\overline{v} = 0$ everywhere. Now, substituting the decomposition in the x-component of the momentum equation (1.3) and averaging yields

$$\overline{u}_t = -(\overline{u'v'})_y \tag{1.48}$$

after some manipulations using $\overline{v} = 0$ and $u'_x + v'_y = 0$. This equation is exact, i.e. it does not depend on small wave amplitudes. It expresses the fact that the zonal mean flow accelerates in response to the convergence of the meridional flux of zonal momentum $\overline{u'v'}$. In regions of constant (but not necessarily zero) flux there is no acceleration.

How does this fit together with our wave solution at $O(a)$? Clearly, the momentum flux $\overline{u'v'}$ is a wave property in that at $O(a^2)$ it can be evaluated from the linear, $O(a)$ wave solution as $\overline{u'_1 v'_1}$. This means we can combine (1.40), which is valid only at $O(a^2)$, with (1.48) to obtain

$$(\overline{u}_2)_t = (\mathsf{p}_2)_t \;\Rightarrow\; \boxed{\overline{u}_t = \mathsf{p}_t + O(a^3)}. \tag{1.49}$$

This innocuous-looking equation is our main result: *to $O(a^2)$ the zonal mean flow acceleration equals the pseudomomentum growth.*

Together with the sign of p that can be read off from (1.36), we see that a growing wave leads to *positive* zonal acceleration where Q_y is negative, i.e. where $U_{yy} > 0$. Indeed, we can re-write (1.49) as

$$\overline{u}_t = \left(\frac{\overline{\eta'^2}}{2}\right)_t \overline{u}_{yy} + O(a^3), \tag{1.50}$$

from which it is easy to see qualitatively that for a growing mode the mean shear is eroded as the mode grows. This is a basis for the sometimes-observed nonlinear growth saturation of marginally unstable modes: the mode grows until the induced mean-flow response shuts off the instability mechanism. At this point the growth ceases and the mode saturates.

2 The beta plane

The waves on shear flows considered above gave us a first example of wave–mean interaction theory. However, it is hard to be more specific without writing down actual wave solutions and those are complicated because $U(y)$ must be non-constant to get waves in the first place. Also, most profiles $U(y)$ have unstable modes, and these will quickly render invalid our linear, $O(a)$ theory for the waves.

For this reason we now turn to fluid systems with a simpler background state that is still sufficient to support waves. The particular example we are going to study is the so-called mid-latitude β-plane, which is a local tangent-plane approximation to our rotating gravitating planet Earth (Pedlosky (1987)).

2.1   The beta-effect

You will know that the Earth spins with frequency Ω around its pole-to-pole axis. If we denote the rotation vector along this axis by $\mathbf{\Omega}$ then we know that the momentum equation relative to the spinning Earth must be augmented by suitable Coriolis and centrifugal forces based on $\mathbf{\Omega}$. The latter can be absorbed in the gravitational potential and need not concern us any further. The former means that a term $\boldsymbol{f} \times \boldsymbol{u}$ must be added to the material derivative, where the Coriolis vector

$$\boldsymbol{f} \equiv 2\mathbf{\Omega}. \tag{2.1}$$

However, it turns out that because of the strong gravitational field of our planet the large-scale motion is mostly "horizontal", i.e. along two-dimensional, nearly spherical stratification surfaces. What is relevant to this horizontal flow is not the full Coriolis vector but only its projection onto the local "upward" direction denoted by the unit vector $\widehat{\boldsymbol{z}}$. That is, if we introduce the usual latitude θ, then we can define the Coriolis *parameter*

$$f \equiv \boldsymbol{f} \cdot \widehat{\boldsymbol{z}} = 2\Omega \sin(\theta). \tag{2.2}$$

This parameter increases monotonically with latitude θ, is zero at the equator, negative in the southern hemisphere, and positive in the northern hemisphere. If we look at a local tangent plane around latitude θ_0 we can introduce the Cartesian coordinates

$$x \qquad \text{Zonal: west-to-east} \tag{2.3}$$
$$y \qquad \text{Meridional: south-to-north} \tag{2.4}$$
$$z \qquad \text{Vertical: low-to-high,} \tag{2.5}$$

and the governing two-dimensional equations are

$$\frac{Du}{Dt} - fv + \frac{1}{\rho}p_x = 0 \tag{2.6}$$
$$\frac{Dv}{Dt} + fu + \frac{1}{\rho}p_y = 0 \tag{2.7}$$
$$u_x + v_y = 0 \tag{2.8}$$

where all fields depend on (x, y, t), as before. The origin of y has been chosen at θ_0 such that $y = R(\theta - \theta_0)$, where $R \approx 6300\text{km}$ is the Earth's radius. Strictly speaking, the tangent-plane approximation is only valid in a range of x and y that is small compared to R, so that the spherical geometry can be neglected. This is often relaxed for the zonal coordinate x, which is usually allowed to go once around the globe such that periodic boundary conditions in x make sense. We continue having solid walls at $y = 0$ and $y = D$. Now, the parameter f is given by a local Taylor expansion as

$$f = 2\Omega \sin(\theta_0) + 2\Omega \cos(\theta_0)(\theta - \theta_0) \equiv f_0 + \beta y, \tag{2.9}$$

which introduces the important constant

$$\beta = \frac{2\Omega}{R} \cos(\theta_0). \tag{2.10}$$

So $\beta > 0$ is the rate of change of f per unit northward distance. This will have a profound dynamical effect, as we shall see.

Can we find a vorticity stream function formulation of (2.6-2.8)? The answer is yes, provided we use the *absolute* vorticity

$$q = v_x - u_y + f_0 + \beta y. \tag{2.11}$$

This is the normal fluid vorticity as seen by a non-rotating, inertial observer: it is the relative vorticity $v_x - u_y$ plus the vorticity due to the rotating frame. It is easy to show that we again get

$$\frac{Dq}{Dt} + q\boldsymbol{\nabla} \cdot \boldsymbol{u} = 0 \quad \Rightarrow \quad \frac{Dq}{Dt} = 0. \tag{2.12}$$

This means that q is a material invariant, just as before. However, the stream function now satisfies

$$\nabla^2\psi + f_0 + \beta y = q, \tag{2.13}$$

which is the new equation. We can note as an aside that if $\beta = 0$ then the constant f_0 can be absorbed in the definition of q, because $q - f_0$ still satisfies (2.12). This means that the $\beta = 0$ dynamics is *exactly* the same as in the non-rotating system studied before! This peculiar fact implies that frame rotation in a two-dimensional incompressible flow is not noticeable.[2]

Let us now consider $O(a)$ waves again. The $O(1)$ background vorticity now satisfies

$$Q = f_0 + \beta y - U_y \quad \Rightarrow \quad Q_y = \beta - U_{yy} \tag{2.14}$$

and this is the only change to the $O(a)$ equations (1.22-1.23). Rayleigh's theorem now states that $\beta - U_{yy}$ must change sign in order to have unstable normal modes. For instance, this means that $|U_{yy}| \leq \beta$ implies stability, i.e. $\beta \neq 0$ has a stabilizing influence.

2.2   Rossby waves

We will now set $U =$const. such that $Q_y = \beta$. Searching for plane-wave solutions $\psi' = \hat\psi \exp(i(kx+ly-\omega t))$ to (1.22-1.23) then yields the Rossby-wave dispersion relation

$$\omega = Uk - \frac{\beta k}{k^2 + l^2} = Uk - \frac{\beta k}{\kappa^2}, \tag{2.15}$$

where κ is the magnitude of the wavenumber vector $\boldsymbol{k} = (k,l)$. The absolute frequency ω is the sum of the Doppler-shifting term Uk and the *intrinsic* frequency

$$\hat\omega = -\frac{\beta k}{\kappa^2} \quad \text{such that} \quad \omega = Uk + \hat\omega. \tag{2.16}$$

The intrinsic frequency captures the wave dynamics relative to the background flow U whereas the absolute frequency ω gives the frequency as seen by an observer fixed on the ground.

We can see by inspection that large-scale Rossby waves (i.e. small κ) have higher intrinsic frequencies than small-scale Rossby waves. The absolute speed of phase propagation for a plane wave at *fixed y* is

$$c \equiv \frac{\omega}{k} = U + \frac{\hat\omega}{k} = U - \frac{\beta}{\kappa^2}. \tag{2.17}$$

We can introduce the intrinsic phase speed $\hat c$ at fixed y such that $c = U + \hat c$:

$$\hat c = \frac{\hat\omega}{k} = -\frac{\beta}{\kappa^2}. \tag{2.18}$$

---

[2]This is not exactly true in the case of unbounded flow: with background rotation there can be pressure-less uniform inertial oscillations $(u,v) = (\cos(f_0 t), -\sin(f_0 t))$, which depend on f_0. You may want to ponder for a second where the loophole in the mathematical argument is that allows this to happen! However, the statement is exact for the bounded channel geometry.

This speed is always negative, i.e. the phase always travels westward relative to the background flow.

The continuity equation for plane waves is $ku' + lv' = 0$, i.e. $\boldsymbol{u}'$ and $\boldsymbol{k}$ are at right angles to each other. This means that Rossby waves are transverse waves, with velocities parallel to lines of constant phase.[3] The absolute Rossby-wave *group velocity* is

$$u_g \equiv \frac{\partial \omega}{\partial k} = U + \beta \frac{k^2 - l^2}{\kappa^4} \tag{2.19}$$

$$v_g \equiv \frac{\partial \omega}{\partial l} = \beta \frac{2kl}{\kappa^4}, \tag{2.20}$$

where the partial derivatives are understood to take (k, l) as independent variables. We recall that, in general, the group velocity gives the speed of propagation for a slowly varying *wavepacket* containing many wave crests and troughs. For example, let the initial condition be

$$\psi'(x, y, 0) = a \exp(-(\boldsymbol{x}\kappa\mu)^2) \exp(i(\boldsymbol{k} \cdot \boldsymbol{x})) \tag{2.21}$$

where $\mu \ll 1$ is a small parameter measuring the scale separation between the wavelength $2\pi/\kappa$ and the scale $1/(\mu\kappa)$ of the slowly varying Gaussian envelope of the wavepacket (the Gaussian is not essential; other smooth envelope functions work as well). Then linear theory predicts that the solution for small enough later times $t \leq O(\mu^{-1})$ is

$$\psi'(x, y, t) = a \exp(-((\boldsymbol{x} - \boldsymbol{c}_g t)\kappa\mu)^2) \exp(i(\boldsymbol{k} \cdot \boldsymbol{x} - \omega t)) \tag{2.22}$$

where $\boldsymbol{c}_g = (u_g, v_g)$ is given by (2.19-2.20). In other words, for $t \leq O(\mu^{-1})$ the wavepacket simply moves with the group velocity. This is long enough to move the envelope a distance comparable to its size. Over longer times, e.g. $t \leq O(\mu^{-2})$, dispersion effects for the wavepacket envelope need to be taken into account, as is done in the (nonlinear) Schrödinger equation for modulated wavepackets, for instance.

We note in passing that a *vectorial* phase velocity $\boldsymbol{c}_p$ can be defined that gives the speed of propagation of individual wave crests or troughs in the xy-plane. It is natural to let $\boldsymbol{c}_p$ be parallel to $\boldsymbol{k}$, which means that the vector

$$\boldsymbol{c}_p = (u_p, v_p) \equiv \frac{\omega}{\kappa^2} \boldsymbol{k} \tag{2.23}$$

is the desired definition. In general, phase and group velocity can have different magnitude as well as direction, which is important when interpreting observations. Finally, we note that the phase velocity c in (2.17) agrees with u_p only if $l = 0$. This is because c is defined as the phase speed at constant y, which as a direction is only parallel to $\boldsymbol{k}$ if $l = 0$.

Indeed, for the special case of $l = 0$ and $U = 0$, the lines of constant phase are $y = $const., the particle velocity satisfies $u' = 0$, the meridional group velocity $v_g = 0$,

---

[3]It is easy to show that this implies that a *single* plane Rossby wave is also an exact solution of the nonlinear equations. This is because the only nonlinearity enters through the nonlinear part of the material derivative, which turns out to be exactly zero for transverse waves.

and the zonal group velocity is

$$u_g = \frac{\beta}{k^2} = -c = -u_p. \tag{2.24}$$

So the group velocity for $l = 0$ is always *eastward* and is opposite to the zonal phase velocity! This enjoyable fact is one of the reasons why atmospheric weather systems usually travel from west to east.

Finally, the *intrinsic* group velocities are defined by replacing ω by $\hat{\omega}$ in (2.19) and (2.23), for example:

$$\hat{u}_g \equiv \frac{\partial \hat{\omega}}{\partial k} = \beta \frac{k^2 - l^2}{\kappa^4} \tag{2.25}$$

$$\hat{v}_g \equiv \frac{\partial \hat{\omega}}{\partial l} = \beta \frac{2kl}{\kappa^4} = v_g. \tag{2.26}$$

We have $(u_g, v_g) = (U, 0) + (\hat{u}_g, \hat{v}_g)$ overall.

2.3   Momentum flux and pseudomomentum

Consider now the wave-induced momentum flux $\overline{u'v'}$, i.e. the meridional flux of x-momentum per unit length across lines of constant y. Using $u' = -v'l/k$ from continuity we get

$$\overline{u'v'} = -\frac{l}{k}\,\overline{v'^2} = -\frac{kl}{k^2}\,\overline{v'^2}, \tag{2.27}$$

which makes it obvious that

$$-\mathrm{sgn}(\overline{u'v'}) = \mathrm{sgn}(kl) = \mathrm{sgn}(v_g), \tag{2.28}$$

provided that $\beta > 0$. This means that a northward-moving wavepacket (i.e. $v_g > 0$) has *negative* momentum flux, and vice versa. Also, (2.28) implies that the lines of constant phase (at fixed time) for a northward moving Rossby wave are slanted from south–east to north–west, and vice versa for a southward moving wave. This can be used to diagnose wave direction from single-time snapshots.

The Rossby-wave pseudomomentum is given by

$$\mathsf{p} = \frac{\overline{\eta'q'}}{2} = -\frac{Q_y}{2}\,\overline{\eta'^2} = -\frac{\beta}{2}\,\overline{\eta'^2} \le 0. \tag{2.29}$$

It is always negative for Rossby waves, which is why in meteorology one sometimes uses the opposite sign convention for p. If we define the wave disturbance energy as

$$E \equiv \frac{1}{2}\left(\overline{u'^2} + \overline{v'^2}\right) \tag{2.30}$$

and make use of

$$\hat{\omega}^2\,\overline{\eta'^2} = \overline{v'^2}, \tag{2.31}$$

which follows directly from (1.25) and $D_t = -i\hat{\omega}$ for plane waves, then it is straightforward to show that

$$\boxed{\mathsf{p} = \frac{k}{\hat{\omega}} E}. \qquad (2.32)$$

This link between p and E for propagating plane waves will be seen to hold much more generally. Among other things, it implies that the sign of p is equal to the sign of the intrinsic zonal phase speed $\hat{c} = \hat{\omega}/k$. It is also easy to show that for a plane wave the momentum flux

$$\overline{u'v'} = \mathsf{p} v_g \qquad (2.33)$$

and therefore

$$\mathsf{p}_t + (\mathsf{p} v_g)_y = 0 \qquad (2.34)$$

holds for slowly varying wavetrains.

Let us now consider the Rossby waves generated by uniform flow U over a small-amplitude undulating sidewall at $y = 0$. That is, the southern channel boundary is undulated at $O(a)$ according to

$$h(x) = h_0 \cos(kx) \qquad (2.35)$$

where $h_0 = O(a)$ and k is the wavenumber of the undulations. (Other shapes of $h(x)$ can be built up by linear superposition of such Fourier modes.) The linear kinematic boundary condition at $y = 0$ becomes

$$v'(x, 0, t) = U h_x = U k h_0 \sin(kx) \quad \text{or more simply} \quad \eta'(x, 0, t) = h_0 \cos(kx) \qquad (2.36)$$

after using (1.25). Now, k is fixed by the undulations and we see from (2.36) that the disturbance is time-independent when observed from the ground. This means that the forced Rossby waves have *absolute* frequency $\omega = 0$ and hence

$$\omega = Uk + \hat{\omega} = 0 \Rightarrow \hat{\omega} = -Uk \Rightarrow U = \frac{\beta}{k^2 + l^2} \Rightarrow l^2 = \frac{\beta}{U} - k^2. \qquad (2.37)$$

This fixes l^2 in terms of k and U. To have propagating waves requires $l^2 > 0$ and we see that this is only possible if U satisfies the so-called Charney–Drazin conditions

$$0 < U < \frac{\beta}{k^2}. \qquad (2.38)$$

If U falls outside this window then $l^2 < 0$ and the waves are trapped, or evanescent, in the y-direction. In other words, in order to excite propagating Rossby waves the background wind must be eastward and not too fast.

It remains to pick the sign of l for propagating waves. If $k > 0$ then we must pick $l > 0$ in order to satisfy the radiation condition $v_g > 0$, i.e. in order to have waves propagating away from the wave maker at $y = 0$. If $k < 0$ the same argument gives $l < 0$. Therefore we have

$$l = \text{sgn}(k) \sqrt{\frac{\beta}{U} - k^2} \qquad (2.39)$$

and the wave field is $\eta' = h_0 \cos(kx + ly)$. The corresponding pseudomomentum from (2.29) is

$$\mathsf{p} = -\frac{\beta}{4}\,h_0^2. \tag{2.40}$$

Interestingly, p does not depend explicitly on either U or k. However, if (2.38) is not satisfied then $\mathsf{p} = 0$. So there is an implicit dependence on Uk^2 here.

Now, we imagine that the wall undulations are growing smoothly from zero to their final amplitude h_0 over some time interval that is long compared to the intrinsic period of the wave. Then there will be a smooth transition zone in space that separates a far-field region without waves from a region with waves. This transition zone, or wave front, will travel with speed $v_g > 0$, which can be computed from (2.20). (For simplicity, we will not consider what happens when the waves reach the other channel wall at $y = D$.) So, below the wave front (2.40) will be valid whilst above it $\mathsf{p} = 0$.

To get to the mean-flow response we note that (1.48) is still valid on the β-plane because $\overline{fu} = f\overline{v} = 0$ and hence we still have

$$\overline{u}_t = -(\overline{u'v'})_y = \mathsf{p}_t + O(a^3). \tag{2.41}$$

Integrating in time we get

$$\overline{u} = U + \mathsf{p} + O(a^3). \tag{2.42}$$

We see that mean-flow acceleration is confined to the wave front, where p changes from zero to $-\beta h_0^2/4$. In other words, the mean flow is *decelerated* as the Rossby waves arrive. Once the wave field is steady there is no further mean-flow change. Furthermore, if we imagine the wall undulations to return to zero again then we see that the mean flow is now *accelerated* back to its original value U. So the waves did not create a lasting, irreversible change in the mean flow. We see from this example that

- only transient waves (i.e. $\mathsf{p}_t \neq 0$) can accelerate the mean flow;
- the mean-flow changes are uniformly bounded (in time) at $O(a^2)$;
- and the mean-flow changes are reversible.

The first point (and its dissipative generalization considered below) is often called a "non-acceleration theorem" in meteorology. The second and third points show that these wave-induced mean-flow changes can be ignored if the amplitude a is small enough. To create lasting mean-flow changes in simple geometry we must allow for wave dissipation.

2.4 Forcing and dissipation

We can include forcing and dissipation by adding a body force $\boldsymbol{F} = (F, G)$ to the right-hand side of the momentum equations:

$$\frac{D\boldsymbol{u}}{Dt} + f\widehat{\boldsymbol{z}} \times \boldsymbol{u} + \frac{1}{\rho}\boldsymbol{\nabla}p = \boldsymbol{F}. \tag{2.43}$$

For instance, the usual Navier–Stokes equations correspond to $\boldsymbol{F} = \nu\nabla^2\boldsymbol{u}$, where $\nu > 0$ is the kinematic viscosity. The forced vorticity equation is

$$\frac{Dq}{Dt} = \widehat{\boldsymbol{z}} \cdot \boldsymbol{\nabla} \times \boldsymbol{F} = G_x - F_y \tag{2.44}$$

and q is not a material invariant any more. We will assume that there is no $O(1)$ part of F, i.e. the background flow is unforced. The expansion of F in terms of wave amplitude a therefore has the form

$$\boldsymbol{F} = \boldsymbol{F}' + O(a^2) \tag{2.45}$$

where $\boldsymbol{F}' = O(a)$. The $O(a)$ vorticity equation is

$$\mathrm{D}_t q' + v' Q_y = \widehat{\boldsymbol{z}} \cdot \boldsymbol{\nabla} \times \boldsymbol{F}' \tag{2.46}$$

but the relation $q' = -Q_y \eta'$ does not hold any more, because of the friction. The most robust definition of pseudomomentum that survives introducing friction is

$$\mathsf{p} \equiv -\frac{\overline{q'^2}}{2Q_y}, \tag{2.47}$$

which equals (1.36) in the absence of friction. The pseudomomentum equation is

$$\mathsf{p}_t + (\overline{u'v'})_y = -\frac{1}{Q_y} \overline{q' \boldsymbol{\nabla} \times \boldsymbol{F}'} \equiv \mathcal{F} \tag{2.48}$$

after introducing the useful short-hand $\mathcal{F}$. Clearly, the total pseudomomentum in the channel is not conserved any more:

$$\frac{\mathrm{d}\mathcal{P}}{\mathrm{d}t} = L \int_0^D \mathcal{F} \, \mathrm{d}y. \tag{2.49}$$

For Rossby waves with constant U the pseudomomentum is always negative, so we see that $\mathcal{F} < 0$ corresponds to wave forcing whereas $\mathcal{F} > 0$ corresponds to wave dissipation.

The forced zonal mean-flow equation at $O(a^2)$ is

$$\overline{u}_t = -(\overline{u'v'})_y + \overline{F} = \mathsf{p}_t - \mathcal{F} + \overline{F} + O(a^3), \tag{2.50}$$

where $\overline{F}$ is evaluated correct to $O(a^2)$.

We now make an important assumption: we assume that $\overline{F}$ can be neglected in (2.50). We will see later that this assumption is linked to the momentum-conservation character of $\boldsymbol{F}$: if the body force does not add any mean zonal momentum to the system then $\overline{F}$ will indeed be negligible in (2.50). (This is physically reasonable in many circumstances; however, it can be difficult to satisfy in numerical simulations. For instance, simple Rayleigh damping $\boldsymbol{F} \propto -(\boldsymbol{u} - \boldsymbol{U})$ will violate this assumption because then $\overline{F}$ turns out to be comparable to $\mathcal{F}$ (Bühler (2000)). With this assumption we obtain the simple result

$$\boxed{\overline{u}_t = \mathsf{p}_t - \mathcal{F} + O(a^3)}, \tag{2.51}$$

which shows that we now have two ways to force the mean flow: either through wave transience or through wave forcing/dissipation.

Let us reconsider the Rossby-wave problem with the undulating wall, but this time with dissipation. We shall assume that

$$\boldsymbol{\nabla} \times \boldsymbol{F}' = -\frac{\alpha}{2} q' \tag{2.52}$$

to $O(a)$ with a constant damping rate $\alpha > 0$. This means that

$$\mathcal{F} = -\frac{\overline{q'\boldsymbol{\nabla} \times \boldsymbol{F}'}}{Q_y} = -\alpha\, \mathsf{p}. \tag{2.53}$$

If the damping rate α is small enough then we are dealing with a slowly varying wavetrain whose structure can be computed by using the plane-wave result (2.33) in (2.48):

$$(\overline{u'v'})_y = (\mathsf{p} v_g)_y = -\alpha\mathsf{p}. \tag{2.54}$$

This is the weak dissipation approximation, which can be formally justified as a first-order approximation in $|\alpha/\hat{\omega}|$. The group velocity v_g is constant in (2.54) and hence we obtain

$$\mathsf{p}(y) = \mathsf{p}(0)\exp\left(-\frac{\alpha}{v_g}y\right) \quad \text{with} \quad \mathsf{p}(0) = -\frac{\beta}{4}h_0^2. \tag{2.55}$$

So the wavetrain amplitude decays exponentially in y with spatial decay rate α/v_g. The mean-flow response to the steady wavetrain is given explicitly by

$$\overline{u}_t = -\mathcal{F} = +\alpha\mathsf{p} = -\alpha\frac{\beta}{4}h_0^2\exp\left(-\frac{\alpha}{v_g}y\right). \tag{2.56}$$

This shows that $\overline{u}_t < 0$ everywhere so the flow decelerates, just as in the spin-up phase of the transient waves. However, unlike the mean-flow response to the transient waves, the dissipative mean-flow changes *persist* after the waves are switched off: dissipative mean-flow changes are irreversible. Connected to this fact is another, even more important fact: the mean-flow change grows without bound in time. Indeed, integrating (2.56) in time gives approximately

$$\overline{u} = U + (1 + \alpha t)\mathsf{p}, \tag{2.57}$$

where the transient non-dissipative acceleration stemming from $\mathsf{p}_t \neq 0$ has been included. The last term gives secular growth, which will invalidate the assumed scaling when $t = O(a^{-2})$. So even small-amplitude waves can give rise to $O(1)$ mean-flow changes after a long time.

Of course, following the solution for such a long time requires a singular perturbation analysis, which is more work than we have done here (our expansion is valid for times $t = O(1)$). However, the conclusion remains valid: wave dissipation can lead to persistent, irreversible mean-flow changes that can grow to $O(1)$ after times $t = O(a^{-2})$.

Let us now consider a variant of this problem with interior wave forcing due to $\boldsymbol{\nabla} \times \boldsymbol{F}'$, say in a region centred around the middle of the channel. The waves propagate away northward and southward and are then subject to dissipation say at a certain distance away from the forcing region. For simplicity, we neglect the influence of the channel walls and we set $U = 0$. By construction we have $\mathcal{F} < 0$ in the wave forcing region near $y = D/2$ and $\mathcal{F} > 0$ is the dissipation regions above and below. If we assume that the wave field is steady on average (i.e. $\mathsf{p}_t = 0$) then the mean-flow response is $\overline{u}_t < 0$ in the dissipation regions, as before. However, we also obtain that $\overline{u}_t > 0$ in the forcing region! This means the mean flow is accelerated eastward in the forcing region, even though we have assumed that $\overline{F} = 0$ and hence there is no net momentum input by the force. How this is possible becomes clear once we consider the wave-induced momentum flux (2.33) above and below the forcing region. Integrate the first equality in (2.50) (with $\overline{F} = 0$) in y to obtain

$$\frac{\mathrm{d}}{\mathrm{d}t} \int_{y_1}^{y_2} \overline{u}\, \mathrm{d}y = -(\overline{u'v'}|_{y_2} - \overline{u'v'}|_{y_1}), \tag{2.58}$$

where $y1$ and $y2$ are southward and northward of the forcing region, respectively. It is easiest to think of these locations to lie in between the forcing region and the dissipation regions. To the north $v_g > 0$ and hence $\overline{u'v'}|_{y_2} < 0$. To the south $v_g < 0$ and hence $\overline{u'v'}|_{y_1} > 0$. This means that there is a net influx of zonal momentum into the region between y_1 and y_2, and (2.58) quantifies how this leads to positive, eastward acceleration.

By momentum conservation the eastward momentum gained by the forcing region is exactly compensated by the westward momentum gained by the dissipation regions. This means that the Rossby waves mediated a **wave-induced non-local momentum transfer** between these regions. The stirring by the body force allowed this momentum transfer to happen, but the forcing itself did not change the momentum in the channel.

It is important to note the signs here: the *arrival and/or dissipation* of Rossby waves accelerates the mean flow westwards, whereas the *departure and/or generation* of Rossby waves leads to eastward acceleration! Such wave-induced momentum transfer between wave forcing and dissipation regions is a key concept in wave–mean interaction theory and atmosphere ocean fluid dynamics. For instance, the Rossby-wave generation caused by large-scale baroclinic instability at mid-latitudes is well known to contribute to the eastward acceleration in the upper troposphere (e.g. Held (2000)).

2.5   Critical layers

We now return to the case of variable $U(y)$. Among other things, understanding this more general case will allow us to understand feedback cycles between wave-induced mean-flow changes and the waves themselves.

To fix ideas we consider again non-dissipative Rossby waves forced by flow past an undulating wall at $y = 0$ (cf. (2.35)), but this time allow for $U(y)$. Specifically, we want to consider the case where $U = 0$ at some $y = y_c$, which is called the *critical line*. The standard approach is to use a normal-mode Ansatz for the steady disturbance stream function

$$\psi' = \hat{\psi}(y) \exp(ikx) \quad \text{such that} \quad q' = \left(\frac{\mathrm{d}^2\hat{\psi}}{\mathrm{d}y^2} - k^2\hat{\psi}\right) \exp(ikx) \tag{2.59}$$

with a modal structure $\hat{\psi}(y)$ to be determined from the relevant $O(a)$ vorticity equation

$$U(y)q'_x + \psi'_x(\beta - U_{yy}) = 0, \text{ or} \tag{2.60}$$

$$U(y)\frac{\mathrm{d}^2\hat{\psi}}{\mathrm{d}y^2} + (\beta - U_{yy} - k^2U)\hat{\psi} = 0. \tag{2.61}$$

This reduces to (2.37) if $U_y = 0$. We see that at y_c the coefficient of the highest derivative in this ODE vanishes, i.e. y_c is a singular point of the equation. Singular points are standard for many linear equations arising in physics, such as Bessel's equation for instance. The standard way to continue is to investigate the power series expansion of U near the singular point in order to find the local behaviour of $\hat{\psi}$ there, which usually involves a logarithmic term centred at y_c. However, we know that the $O(a)$ equations are only a small-amplitude approximation to the nonlinear fluids equations. Linear critical layer theory might be a poor guide to real fluid dynamics! Indeed, most linear (or weakly nonlinear) critical layer theory is wholly unrealistic, as we shall see that the flow at a critical layer is almost always *strongly* nonlinear. This is an important (and not very widely appreciated) fact for practical purposes, which puts a lot of theoretical critical layer research into perspective.

To see the breakdown of linear theory at the critical layer we shall use a ray-tracing (or JWKB) approximation valid for slowly varying waves on a slowly varying shear flow $U(y)$ (e.g. Lighthill (1978)). This is a very powerful general tool in wave theory. It amounts to approximating the wave structure everywhere by a *local* plane wave determined by the local value of $U(y)$. This means making the ray-tracing Ansatz

$$\psi' = \tilde{\psi}(y)\exp(ikx)\exp\left(i\int_0^y l(\bar{y})\mathrm{d}\bar{y}\right) \quad \text{with} \tag{2.62}$$

$$l(y) = +\sqrt{\frac{\beta}{U(y)} - k^2} \tag{2.63}$$

for the steady stream function. So the local meridional wavenumber $l(y)$ itself varies with y in according to the plane-wave formula (2.39). The zonal wavenumber $k > 0$ does not change. The amplitude $\tilde{\psi}$ is undetermined at this stage. For this ray-tracing approach to make sense we need $\beta \gg |U_{yy}|$ and also a condition such as

$$\left(\frac{U_y}{U}\right)^2 \ll l^2 \tag{2.64}$$

that quantifies that $U(y)$ is slowly varying over a meridional wavelength $2\pi/l$. (The exact asymptotic conditions for ray-tracing to be accurate are more complicated than (2.64), but (2.64) is a good guide in practice.)

Now, we see immediately from (2.63) that at the critical line y_c the local wavenumber $l \to \infty$, so the wave structure becomes non-smooth near y_c. Specifically, if U is proportional to $y_c - y$ just below the critical line then

$$l \propto (y_c - y)^{-1/2} \tag{2.65}$$

there. We can consider the motion of a wave front moving with v_g in order to find out how long it takes for a wave front to reach the critical line y_c. Using ray-tracing we know that v_g will depend on y through $l(y)$ such that

$$v_g = \beta\frac{2kl}{(k^2 + l^2)^2} \propto \frac{1}{l^3} \propto (y_c - y)^{3/2} \tag{2.66}$$

just below y_c. This means that $v_g \to 0$ at y_c and so the wave front slows down. The travel time near the critical line can be easily estimated from

$$\int dt = \int \frac{dy}{v_g} \propto \int \frac{dy}{(y_c - y)^{3/2}} \propto (y_c - y)^{-1/2} + \text{const.}, \qquad (2.67)$$

which shows that it takes infinite time for the wave front to reach the critical line $y = y_c$.

This strongly suggests that the wavetrain accumulates just below y_c, in a region that has been named the critical *layer*. Now, the wave amplitude (i.e. $|\tilde{\psi}|$) for a slowly varying wavetrain is determined by the steady pseudomomentum conservation law

$$(\overline{u'v'})_y = (\mathsf{p}v_g)_y = 0 \Rightarrow \mathsf{p}(y) = \mathsf{p}(0) \frac{v_g(0)}{v_g(y)}, \qquad (2.68)$$

which shows that p also becomes unbounded. A good physical definition for the local wave amplitude is the non-dimensional overturning amplitude

$$a \equiv \max |\eta'_y|, \qquad (2.69)$$

where the maximum is taking over the x variable. This is because the undulating vorticity contours are given by

$$
\begin{aligned}
q &= Q + q' + O(a^2) & (2.70) \\
&= f_0 + \beta y - \beta \eta' + O(a^2) & (2.71) \\
q_y &= \beta(1 - \eta'_y) + O(a^2), & (2.72)
\end{aligned}
$$

which shows that overturning (i.e. $q_y = 0$) occurs first where η'_y exceeds unity. Such overturning means that $a \approx 1$ and linear theory must fail. With this amplitude definition we have the ray-tracing relations

$$a^2 = 2 \overline{\eta'^2_y} = 2l^2 \overline{\eta'^2} = -\frac{4l^2}{\beta} \mathsf{p}. \qquad (2.73)$$

In the critical layer this implies

$$a^2 \propto \frac{l^2}{v_g} \propto l^5 \propto U^{-5/2} \Rightarrow a \propto U^{-5/4} \propto (y_c - y)^{-5/4}, \qquad (2.74)$$

and hence waves must *break nonlinearly* in the critical layer just below y_c.

In summary, we have shown that linear theory must break down in the critical layer just below the critical line y_c. Ray-tracing itself also breaks down there, as can be checked from (2.64), which is violated in the critical layer. Furthermore, even the assumption of a steady state has broken down, although we could only formulate this assumption within linear theory. So, linear theory has predicted its own comprehensive breakdown in the critical layer. This is actually a good, strong scientific result, much preferable to the alternative of having a theory that continues to work fine even though it has ceased to be valid!

What happens at a true critical layer can be elucidated by more sophisticated theory as well as by nonlinear numerical simulations. The picture that emerges is roughly as follows (e.g Haynes (2003), Haynes (1985)).

- As the wave gets close to the critical line the overturning amplitude grows and eventually closed stream lines are formed in the critical layer, giving the flow a characteristic "cat eyes" appearance;
- the flow becomes unstable and breaks down into strongly nonlinear two-dimensional turbulence;
- the turbulence mixes the fluid and the materially advected vorticity distribution becomes approximately uniform, or *homogenized* in the critical layer (eventually, this is due to viscous diffusion acting on small-scale vorticity generated by the turbulence).

Surprisingly, we can compute the mean-flow changes in the critical layer based only on the homogenization of vorticity. If $\nabla q = 0$ then

$$\overline{q}_y = 0 = \beta - \overline{u}_{yy} \Rightarrow \overline{u}(y) = A + By + \frac{\beta y^2}{2} \qquad (2.75)$$

must hold in the critical layer for some A, B. If $\overline{u}$ should match $U(y)$ outside the critical layer then it is not hard to see that the mean-flow change $\overline{u} - U$ is a quadratic centred at the middle of the critical layer and zero at the edges of this layer. The sign of $\overline{u} - U$ is always negative and hence the net mean-flow acceleration has been westward, the same as in the case of wave dissipation. So we can see that wave breaking is similar to wave dissipation in some respects: it destroys the wave and it leads to westward mean-flow acceleration. However, once q has been homogenized in the critical layer then there is no further mean-flow acceleration possible inside it. This is different from straightforward wave dissipation and this fact goes hand-in-hand with the finding that mature critical layers tend to reflect rather than absorb further incoming Rossby waves: they cannot absorb any more momentum flux and hence must reflect the incoming waves (Killworth and McIntyre (1985)).

It is noteworthy that linear theory breaks down in the critical layer but can still be used to compute the momentum flux $\overline{u'v'}$ outside from the critical layer. Momentum conservation and the associated momentum fluxes are fully nonlinear concepts that hold without restriction to small wave amplitudes and this is another way of seeing that the mean-flow acceleration had to be westward.

In summary, we have seen that variable $U(y)$ can lead to critical lines where $U = 0$ and that wave propagation is not possible past such lines. Instead, nonlinear wave breaking occurs in a critical layer just below the critical line. This involves overturning of the vorticity contours and eventual homogenization of the vorticity field, concomitant with westward acceleration of the mean flow. It is straightforward to show that all of the above applies equally well for Rossby waves with $c \neq 0$, which encounter critical lines where $U = c$. In general, critical lines occur where the *intrinsic* phase speed $\hat{c} = c - U$ is zero.

2.6 Reflection

Variable $U(y)$ can lead to a second strong effect: wave reflection. This occurs when $U(y)$ reaches the upper speed limit β/k^2 and hence $l = 0$ there, from (2.63). This implies that $v_g = 0$ there as well and this means that simple ray-tracing predicts unbounded amplitudes again, just as in the critical layer case. However, this time

$$v_g^{-1} \propto l^{-1} \propto (\beta/k^2 - U)^{-1/2} \tag{2.76}$$

near the reflection line where $U = \beta/k^2$. This implies that the wave front reaches the reflection line in finite time. What actually happens is that the wave is reflected at this location, which is an effect that is not captured by simple ray-tracing. The wave field then consists of an two waves, one going northward and the reflected one going southward. Eventually, a steady state establishes itself that has nonzero pseudomomentum p(y) between the southern wall and the reflection line and zero p(y) to the north of the reflection line. The pseudomomentum or momentum flux $\overline{u'v'} = 0$ everywhere.

Wave reflection means that simple ray tracing breaks down (cf. (2.64)), but linear theory remains valid. Indeed, the modal equation (2.60) has no singularity at the reflection line, in contrast to the situation at the critical line, and it can straightforwardly be solved locally in terms of Airy functions. (As an exercise, you may consider which parts of the above remain true if it should happen that $U_y = 0$ at the reflection line.)

2.7 Another view on mean-flow acceleration

Recall the exact mean-flow equation (without forcing)

$$\overline{u}_t + (\overline{u'v'})_y = 0. \tag{2.77}$$

Using the equally exact definition of $q' = v'_x - u'_y$ and that $u'_x + v'_y = 0$ we can easily prove the *Taylor identity*

$$(\overline{u'v'})_y = -\overline{q'v'}. \tag{2.78}$$

Doing this mirrors the steps below (1.37), which made no use of small-amplitude approximations. Now, this means that (2.77) can be re-written as

$$\overline{u}_t = \overline{q'v'}. \tag{2.79}$$

In this form the exact zonal mean-flow acceleration equals the northward flux of vorticity. Now, using the small-amplitude results $q' = -Q_y\eta'$ and $D_t\eta' = v'$ we see that

$$\overline{u}_t = -Q_y \overline{\eta' D_t \eta'} = \frac{\partial}{\partial t}\left(-Q_y \frac{\overline{\eta'^2}}{2}\right) = \mathsf{p}_t \tag{2.80}$$

holds to $O(a^2)$. This slick short derivation of our main result is worth knowing.

The exact law (2.79) also allows understanding why growing Rossby waves must be accompanied by westward acceleration, as follows. The absolute vorticity q is materially advected and increases with y in the undisturbed reference configuration. Consider one particular contour of q, say the one at $y = 5$ in the reference configuration, on which $q = Q(5)$. Now, growing Rossby waves undulate the contours of constant q such that the area between our contour and the line $y = 5$ is filled with $q < Q(5)$ where our contour is *northward* of $y = 5$ or with $q > Q(5)$ where our contour is *southward* of it. This means that there has been a net negative transport of vorticity across the line $y = 5$, i.e. $\overline{q'v'} < 0$

and hence $\bar{u}_t < 0$ follows. This qualitative picture remains valid nonlinearly, i.e. it does not depend on small wave amplitudes: the westward acceleration due to the arrival of Rossby waves is a robust feature.

3 Internal gravity waves

We now turn to a new physical system, with a new kind of waves (e.g. Staquet and Sommeria (2002)). Doing this will get us one step closer towards a robust theory of wave–mean interactions, i.e. a theory that is informed by as many different physical applications as possible. Internal gravity waves are important in their own right as contributing significantly to the global circulation of the atmosphere and oceans. They are usually too small in spatial scale (especially vertical scale) to be resolved in global computer models and this makes their theoretical study especially important, as global models depend crucially on the theory.

3.1 Boussinesq equations

The two-dimensional ideal Boussinesq equations are

$$\frac{Du}{Dt} + \frac{1}{\rho_0}p_x = 0 \tag{3.1}$$

$$\frac{Dw}{Dt} + \frac{1}{\rho_0}p_z = b \tag{3.2}$$

$$\frac{Db}{Dt} + N^2 w = 0 \tag{3.3}$$

$$u_x + w_z = 0. \tag{3.4}$$

Here we work in the xz-plane, with x west–east as before and z being altitude. The two-dimensional velocity $\boldsymbol{u} = (u, w)$ is non-divergent and the density ρ_0 is a uniform constant, as before. The new variable is the *buoyancy* $b(x, z, t) = -g(\rho - \rho_0)/\rho_0$, which arises due to joint effect of gravity and density contrasts in the vertical (e.g. Pedlosky (1987)). This leads to a force b in the vertical, as indicated. This force is upward if there is positive buoyancy $b > 0$ and vice versa.

The evolution equation for b expresses the fact that there are material *stratification surfaces* (or lines in two dimensions)

$$\theta = b + N^2 z \quad \text{such that} \quad \frac{D\theta}{Dt} = 0 \tag{3.5}$$

and at undisturbed rest we have $b = 0$ and linear stratification $\theta = N^2 z$. Here the constant N is the *buoyancy frequency*, for reasons to become clear below. In the ocean these stratification surfaces are surfaces of constant density (called "isopycnals"), whereas in the atmosphere they are surfaces of constant entropy ("isentropes"). The Boussinesq equations are a useful approximation in both cases, although in the atmospheric case the vertical extent of the domain must be small compared to a density scale height (approx 7km), so that we can neglect the density decay in ρ_0.

We can eliminate the pressure p as before by focusing on the vorticity equation, which is

$$\frac{D}{Dt}(u_z - w_x) = -b_x. \tag{3.6}$$

The nonzero right-hand side shows how y-vorticity can be generated by sloping stratification surfaces: this is called *baroclinic generation of vorticity*. You can convince yourself that the sign of the vorticity generation is such that sloping stratification surfaces tend to always rotate back towards the horizontal. Due to inertia, they overshoot the horizontal equilibrium position and this is the basic oscillation mechanism for internal gravity waves.

3.2 Linear gravity waves

Consider small-amplitude waves relative to a steady background state with $b = 0$ and $\boldsymbol{u} = (U(z), 0)$. The $O(a)$ equations are

$$\begin{aligned}
D_t(u'_z - w'_x) + b'_x &= 0 \tag{3.7} \\
D_t b' + N^2 w' &= 0 \tag{3.8} \\
u'_x + w'_z &= 0. \tag{3.9}
\end{aligned}$$

Introducing a stream function ψ' such that $u' = +\psi'_z$ and $w' = -\psi'_x$ gives

$$D_t(\psi'_{xx} + \psi'_{zz}) + b'_x = 0 \tag{3.10}$$

and after cross-differentiation we can eliminate b' to obtain the single equation

$$\boxed{D_t D_t(\psi'_{xx} + \psi'_{zz}) + N^2 \psi'_{xx} = 0} \tag{3.11}$$

This is called the *Taylor–Goldstein* equation. If $U = \text{const.}$ we can look for plane-wave solutions $\psi' = \hat{\psi} \exp(i(kx + mz - \omega t))$ in terms of absolute frequency ω and wavenumber vector $\boldsymbol{k} = (k, m)$. As before, the intrinsic frequency $\hat{\omega} = \omega - Uk$ and we have the usual plane-wave relations

$$\frac{\partial}{\partial t} = -i\omega, \quad \frac{\partial}{\partial x} = ik, \quad \frac{\partial}{\partial z} = im, \quad \Rightarrow \quad D_t = -i\hat{\omega}. \tag{3.12}$$

Substituting in (3.11) gives

$$-\hat{\omega}^2(-k^2 - m^2) - k^2 N^2 = 0 \quad \Rightarrow \quad \boxed{\hat{\omega}^2 = N^2 \frac{k^2}{k^2 + m^2}}, \tag{3.13}$$

which is the dispersion relation for internal gravity waves. It is one of the most peculiar dispersion relations found in the natural world. We note a number of important facts about gravity waves.

1. There are two roots $\hat{\omega}$ for each $\boldsymbol{k}$:

$$\hat{\omega} = \pm N \frac{k}{\kappa}, \tag{3.14}$$

where $\kappa = |\mathbf{k}|$. This means there are two independent wave modes for each $\mathbf{k}$, which is because we need to specify initial conditions for *two* fields in the linear Boussinesq equations, say for the stream function ψ' and for the buoyancy b'. This is unlike the Rossby-wave case, where there was only one mode and only one field, say q'.

2. The continuity equation implies $\mathbf{u}' \cdot \mathbf{k} = 0$ and hence gravity wave are transverse waves, with fluid velocities perpendicular to $\mathbf{k}$ and hence tangent to planes of constant phase. The same was true for Rossby waves, and single plane gravity waves are again exact solutions of the nonlinear equations.

3. There is a finite frequency bandwidth

$$0 \le \hat{\omega}^2 \le N^2. \tag{3.15}$$

The lower limit is attained when $k = 0$ and the flow is entirely horizontal. The upper limit is attained when $m = 0$ and the flow entirely vertical, with bands of fluid moving up and down with horizontal spacing $2\pi/k$. These are called buoyancy oscillations, which explains the name of N. This most rapid gravity wave has a frequency of about 7 minutes in the atmosphere and about 1 hour in the ocean.

4. The frequency does not depend on spatial scale, i.e. $\hat{\omega}$ is a function only of k/m. Specifically, if polar coordinates are used for the wavenumber vector such that $k = \kappa \cos \alpha$ and $m = \kappa \sin \alpha$ then we have

$$\hat{\omega}^2 = N^2 \cos^2 \alpha. \tag{3.16}$$

So the frequency depends only on the angle of $\mathbf{k}$ but not on its magnitude.

5. The intrinsic group velocities are

$$\boxed{\hat{u}_g = \frac{\partial \hat{\omega}}{\partial k} = \pm N \frac{m^2}{\kappa^3}, \quad \hat{w}_g = \frac{\partial \hat{\omega}}{\partial m} = \mp N \frac{km}{\kappa^3}}, \tag{3.17}$$

where the sign cases correspond to the two wave modes. This means that $k\hat{u}_g + m\hat{w}_g = 0$, which is a direct consequence of (3.16). Therefore, just like the particle velocity, the group velocity is also perpendicular to $\mathbf{k}$ and hence the intrinsic group velocity *makes a right angle* with the intrinsic phase velocity (cf. 2.23 with ω replaced by $\hat{\omega}$). Furthermore, it is easy to see that for *both* wave modes

$$\operatorname{sgn}(\hat{\omega}/k) = \operatorname{sgn}(\hat{u}_g), \quad \text{but} \quad \operatorname{sgn}(\hat{\omega}/m) = -\operatorname{sgn}(\hat{w}_g). \tag{3.18}$$

This means that a wave with eastward intrinsic phase speed (i.e. $\hat{c} = \hat{\omega}/k > 0$) also has an eastward intrinsic group velocity. On the other hand, *downward* phase propagation corresponds to *upward* group velocity! Historically, this has been very important for the correct interpretation of observations (Hines (1989)).

6. The wave-induced vertical flux of zonal momentum is

$$\overline{u'w'} = -\frac{k}{m}\,\overline{u'^2} = -\frac{m}{k}\,\overline{w'^2} \tag{3.19}$$

after using the continuity equation, and this means that its sign is given by

$$\operatorname{sgn}(\overline{u'w'}) = -\operatorname{sgn}(km) = \operatorname{sgn}(\hat{u}_g)\operatorname{sgn}(\hat{w}_g) \tag{3.20}$$

for both wave modes. So, an upward–eastward-moving gravity wave has a positive vertical flux of zonal momentum and an upward–westward-moving gravity wave has a negative flux.

3.3   Gravity wave pseudomomentum

Now we would like to find the pseudomomentum of gravity waves. We work in analogy with the Rossby-wave case. First, we define the vertical particle displacement ζ' to $O(a)$ by

$$D_t\zeta' = w' \quad \Rightarrow \quad b' = -\zeta' N^2 \tag{3.21}$$

follows from (3.8). We now multiply (3.7) by ζ', average over x, and manipulate the terms:

$$\overline{\zeta' D_t(u'_z - w'_x)} = -N^2\overline{\zeta' b'_x} \quad = \quad \overline{b' b'_x} = \frac{1}{2}\overline{(b'^2)}_x = 0 \tag{3.22}$$

$$\Rightarrow D_t\overline{\zeta'(u'_z - w'_x)} - \overline{w'(u'_z - w'_x)} \quad = \quad 0 \tag{3.23}$$

$$\Rightarrow D_t\overline{\zeta'(u'_z - w'_x)} = \overline{w'u'_z} \quad = \quad (\overline{u'w'})_z - \overline{u'w'_z} = (\overline{u'w'})_z. \tag{3.24}$$

We have made repeated use of $u'_x + w'_z = 0$ here. We now define the gravity-wave pseudomomentum

$$\boxed{\mathsf{p} = -\overline{\zeta'(u'_z - w'_x)} = \frac{1}{N^2}\overline{b'(u'_z - w'_x)} \quad \Rightarrow \quad \mathsf{p}_t + (\overline{u'w'})_z = 0}\ . \tag{3.25}$$

This can be compared to the Rossby-wave pseudomomentum (2.29). The most conspicuous difference is a factor of two, which is related to the fact that the wave energy for gravity waves is

$$E = \frac{1}{2}\left(\overline{u'^2} + \overline{w'^2} + \frac{\overline{b'^2}}{N^2}\right), \tag{3.26}$$

which contains a second, potential energy term. Plane gravity waves (in the absence of background rotation) obey energy equipartition and hence $E = \overline{b'^2}/N^2$ holds, i.e. the total wave energy is twice the kinetic energy. This means that the generic expression $\mathsf{p} = kE/\hat{\omega} = E/\hat{c}$ holds for plane gravity waves as well. This is an easy way to see that eastward waves have positive pseudomomentum and vice versa. In addition, for plane waves the pseudomomentum flux $\overline{u'w'} = \mathsf{p}w_g$.

3.4   Another view on mean-flow acceleration, again

Just as in the Rossby-wave case there is a short-cut to the main result for gravity waves. Recall that the exact mean-flow equation (without forcing) can be rewritten

$$\overline{u}_t = -(\overline{u'w'})_z = -\overline{(u'_z - w'_x)w'} \tag{3.27}$$

by using the Taylor identity. Now, using the small-amplitude results $D_t(u'_z - w'_x) = -b'_x$, $D_t\zeta' = w'$, and $\zeta' = -b'/N^2$ we find that

$$\overline{u}_t = D_t\left(-\overline{(u'_z - w'_x)\zeta'}\right) - \overline{\zeta' b'_x} = \mathsf{p}_t - 0 \tag{3.28}$$

holds to $O(a^2)$. Thus we again have a quick route to the pseudomomentum definition as well as to the mean-flow acceleration equation.

3.5 Rossby versus gravity waves

Here is a brief summary of some aspects of two-dimensional Rossby and non-rotating gravity waves. There are similarities but the two are certainly not identical.

	Rossby waves	Gravity waves
Domain	(x, y)	(x, z)
Material invariant	absolute vorticity	stratification
	$q = v_x - u_y + f_0 + \beta y$	$\theta = b + N^2 z$
Pseudomomentum	$\mathsf{p} = -\overline{q'^2}/(2\beta)$	$\mathsf{p} = \overline{b'(u'_z - w'_x)}/N^2$
	sign-definite	not sign-definite
Momentum flux	$\overline{u'v'}$	$\overline{u'w'}$
Dispersion relation	$\hat{\omega} = -\beta k/(k^2 + l^2)$	$\hat{\omega} = \pm N k/\sqrt{k^2 + l^2}$

Both waves can be viewed as undulations of the contours marking constant values of the respective material invariant. There has to be a background gradient in this invariant (i.e. nonzero β and $N^2 > 0$) for there to be a restoring mechanism and hence waves. Single plane waves always satisfy $\mathsf{p} = kE/\hat{\omega}$ and the corresponding pseudomomentum fluxes are $\overline{u'v'} = \mathsf{p}v_g$ and $\overline{u'w'} = \mathsf{p}w_g$, respectively.

3.6 Mountain waves

Consider a uniform background flow with velocity $U > 0$ over a mountain described by a small-amplitude surface undulation

$$h(x) = h_0 \cos(kx), \quad h_0 = O(a) \tag{3.29}$$

such that the vertical particle displacement at the lower boundary $z = 0$ is given by $\zeta'(x, 0) = h(x)$. The wave is steady with respect to the ground and hence the absolute frequency $\omega = 0$. This implies

$$0 = \omega = \hat{\omega} + Uk \quad \Rightarrow \quad \frac{\hat{\omega}}{k} = \hat{c} = -U \quad \Rightarrow \quad U = \frac{N}{\kappa}, \tag{3.30}$$

where the last expression uses (3.14) and the mode selection is dictated by the sign of U. For $U > 0$ the *lower* wave mode branch had to be selected to satisfy $\omega = 0$ and if $U < 0$ then the upper branch would have been selected. We have assumed $U > 0$ and hence we will use the lower sign in all expressions for the group velocities etc. The horizontal wavenumber k is given and we solve the third expression in (3.30) for the vertical wavenumber m to get $m^2 = N^2/U^2 - k^2$. The vertical group velocity $\hat{w}_g = w_g$ must be positive for waves propagating away from the lower boundary and hence $km > 0$. A convenient sign convention consistent with this and the mode selection is

$$k > 0, \quad m > 0, \quad \hat{\omega} < 0. \tag{3.31}$$

Flipping *all* of these signs would work equally well as a sign convention.

This means we have

$$m = +\sqrt{\frac{N^2}{U^2} - k^2},$$ (3.32)

and the window for propagating waves is $0 < U < N/k$. (Repeating the argument for $U < 0$ yields the window $0 < U^2 < N^2/k^2$, i.e. positive and negative wind speeds are allowed, unlike in the Rossby-wave case.) The vertical group velocity is

$$w_g = \frac{kmN}{\kappa^3} = \frac{kU^2}{N}\sqrt{1 - \frac{k^2 U^2}{N^2}}.$$ (3.33)

Now, the linear fields inside the established wavetrain are easily found to be

$$\zeta'(x,z) = -b'(x,z)/N^2 \quad = \quad h_0 \cos(kx + mz)$$ (3.34)

$$u'(x,z) \quad = \quad Umh_0 \sin(kx + mz)$$ (3.35)

$$w'(x,z) \quad = \quad -Ukh_0 \sin(kx + mz)$$ (3.36)

$$\mathsf{p} = \overline{b'(u'_z - w'_x)}/N^2 \quad = \quad -Uh_0^2\kappa^2 \overline{\cos^2(kx + mz)}$$ (3.37)

$$\Rightarrow \mathsf{p} \quad = \quad -h_0^2\frac{U\kappa^2}{2} = -\frac{h_0^2 N^2}{2U},$$ (3.38)

where m is given by (3.32) and the last line made use of $U = N/\kappa$. The expression for p is easily checked to be consistent with $\mathsf{p} = E/\hat{c}$. It is interesting to note that p diverges as $U \to 0+$, even though $\mathsf{p} = 0$ for $U = 0$.

The drag force D exerted on the mountain range is equal to minus the wave-induced momentum flux $\overline{u'w'}$. From the linear solution this yields

$$\boxed{D = -\overline{u'w'} = -\mathsf{p}w_g = +\frac{h_0^2 k}{2}U\sqrt{N^2 - k^2 U^2}}.$$ (3.39)

So the mountain range feels a net force in the x-direction of this magnitude. The drag is linear in U for small U and, unlike p, goes to zero at the speed limits $U = 0$ and $U = N/k$. It is maximal for $U_* = N/(k\sqrt{2})$, with value $D_* = h_0^2 N^2/4$.

The exact zonal mean-flow acceleration equation for the Boussinesq system is

$$\overline{u}_t + (\overline{u'w'})_z = 0,$$ (3.40)

which yields

$$\overline{u}_t = \mathsf{p}_t + O(a^3)$$ (3.41)

as in the Rossby-wave case. Therefore the mean flow inside the wavetrain is changed to

$$\overline{u} = U - \frac{h_0^2 N^2}{2U}$$ (3.42)

as the wavetrain arrives. This nonlocal momentum transfer precisely balances the equal and opposite drag on the mountain below.

3.7   Forcing and dissipation

To effect lasting mean-flow changes we again have to allow for wave dissipation. The simplest way to do this is to add a body force $\boldsymbol{F} = (F, G)$ and a heating term H to the right-hand sides of the Boussinesq equations. Assuming that these terms have no $O(1)$, background part, we obtain for the linear equations

$$D_t(u'_z - w'_x) + b'_x = F'_z - G'_x \tag{3.43}$$

$$D_t b' + N^2 w' = H' \tag{3.44}$$

$$\Rightarrow \quad \mathsf{p}_t + (\overline{u'w'})_z = \frac{1}{N^2}\overline{H'(u'_z - w'_x)} + \frac{1}{N^2}\overline{b'(F'_z - G'_x)} \equiv \mathcal{F} \tag{3.45}$$

where $\mathsf{p} = \overline{b'(u'_z - w'_x)}/N^2$. If we assume again that $\overline{F} = 0$ at $O(a^2)$, then for a steady wavetrain we obtain the mean-flow acceleration equation

$$\bar{u}_t = -(\overline{u'w'})_z = -\mathcal{F} + O(a^3). \tag{3.46}$$

The simplest case has no body force but includes radiative damping given by the *Newtonian cooling* approximation

$$H' = -\alpha\, b', \quad \Rightarrow \quad \mathcal{F} = -\alpha\mathsf{p}, \quad \Rightarrow \quad \boxed{\bar{u}_t = +\alpha\mathsf{p}}, \tag{3.47}$$

where $\alpha > 0$ is a constant. In the atmosphere, this kind of damping arises due to radiative energy transfers within the atmosphere, which tend to dampen temperature disturbances. As the sign of p is the sign of $\hat{c}$, we see that the dissipation of left-going intrinsic wave speeds leads to negative $\bar{u}_t$ and vice versa.

The wave structure is found from the steady pseudomomentum equation (3.45) and $\mathsf{p}w_g = \overline{u'w'}$ as follows:

$$(\overline{u'w'})_z = -\alpha\mathsf{p} = -\frac{\alpha}{w_g}\overline{u'w'} \tag{3.48}$$

$$\Rightarrow \quad \overline{u'w'}(z) = \overline{u'w'}(0)\exp(-\alpha z/w_g) \tag{3.49}$$

$$\Rightarrow \quad \mathsf{p}(z) = \mathsf{p}(0)\exp(-\alpha z/w_g), \tag{3.50}$$

which together with (3.47) gives the mean-flow acceleration profile. Here, the ground-level $\mathsf{p}(0) = -h_0^2 N^2/(2U)$, as computed before.

3.8   Critical layers

We turn to the case of a slowly varying $U(z)$ and focus attention at the location of a critical line z_c such that $U(z_c) = 0$. We assume again that a ray-tracing, or JWKB approximation, is feasible, i.e. we look for a locally plane-wave solution with $k > 0$ and $\omega = 0$ constant but with variable $m(z)$ such that (3.30) is satisfied for all z. From (3.32) we anticipate that m will become very large near the critical line and we will for simplicity from now on assume that $m^2 \gg k^2$ everywhere. Gravity waves obeying this scaling are called *hydrostatic* gravity waves and for mountain waves this corresponds to $U \ll N/k$.

For hydrostatic mountain gravity waves we simply have

$$m(z) = \frac{N}{U(z)} \quad \text{and} \quad w_g(z) = \frac{kU^2}{N}. \tag{3.51}$$

This shows that $m \to \infty$ and $w_g \to 0$ as $z \to z_c$. The JWKB approximation will be valid if

$$m^2 \gg \left(\frac{U_z}{U}\right)^2 \quad \Rightarrow \quad \frac{N^2}{(U_z)^2} \gg 1. \tag{3.52}$$

The non-dimensional parameter $N^2/(U_z)^2$ is called the *Richardson* number and it is typically larger than unity in the atmosphere or oceans. Interestingly, we see that it is possible to have a gravity-wave critical layer at which JWKB theory remains valid, i.e. there are infinitely many wave oscillations below the critical line. This was not so in the Rossby-wave case. (A more detailed investigation in fact shows that the criterion is $N^2/(U_z)^2 > 1/4$, which is very mild and coincides with the linear stability criterion of the background shear flow.) It is easy to show that the travel time to the critical layer is again infinite.

The wave structure follows from (3.48) with the important difference that w_g is now a function of z. This yields

$$\overline{u'w'}(z) = \overline{u'w'}(0) \exp\left(-\alpha \int_0^z \frac{d\bar{z}}{w_g(\bar{z})}\right) = \overline{u'w'}(0) \exp\left(-\frac{\alpha N}{k} \int_0^z \frac{d\bar{z}}{U^2(\bar{z})}\right). \tag{3.53}$$

This shows that dissipation per unit vertical length is enhanced in regions where U^2 is small. If U is smooth at z_c then the integral diverges as $z \to z_c$ and we obtain that $\overline{u'w'}(z_c) = 0$, i.e. the wave has been completely absorbed in a critical layer just below the critical line. The mean-flow acceleration profile follows as

$$\bar{u}_t = \alpha \mathrm{p} = \frac{\alpha}{w_g} \overline{u'w'}(z) = \frac{\alpha N}{kU^2} \overline{u'w'}(0) \exp\left(-\frac{\alpha N}{k} \int_0^z \frac{d\bar{z}}{U^2(\bar{z})}\right) \tag{3.54}$$

$$\Rightarrow \quad \bar{u}_t = -\frac{\alpha h_0^2 N^2}{2U} \exp\left(-\frac{\alpha N}{k} \int_0^z \frac{d\bar{z}}{U^2(\bar{z})}\right), \tag{3.55}$$

where the second form uses (3.39).

Whilst the JWKB approximation is valid in the critical layer we can not be assured that linear theory itself will be valid there. In fact, it is often not. To test for validity of linear theory we once again consider an overturning amplitude $a(z)$ such that

$$a(z) = \max_x |\zeta_z'| = \max_x |b_z'/N^2|. \tag{3.56}$$

If $a = 1$ then the stratification surfaces $\theta = N^2 z + b'$ overturn, i.e. $\theta_z = 0$ at some location, and linear theory must break down. Indeed, linear theory is based on $a \ll 1$. For sinusoidal waves it follows that

$$a^2 = 2m^2 \overline{\zeta'^2} = 2\frac{m^2}{\hat{\omega}^2} \overline{w'^2} = -2\frac{km}{\hat{\omega}^2} \overline{u'w'} = -2\frac{N}{kU^3} \overline{u'w'}. \tag{3.57}$$

Combined with (3.53) this indicates that a^2 will eventually go to zero at the critical line, but before that it will make an excursion to large values due to the occurrence of U^3 in the denominator. This shows that a^2 is less well behaved than $\overline{u'w'}$. Whether this excursion of a^2 to larger values will lead to $a \approx 1$ and hence to the breakdown of linear theory depends on the detailed circumstances. However, it is believed that in practice most gravity waves will break nonlinearly before they reach the critical line.

We have focused on mountain waves, which have zero absolute frequency ω. However, the above computations are easily adapted for waves with $\omega \neq 0$. Specifically, waves with phase speed $c = \omega/k$ will encounter critical layers where $U = c$, i.e. where the intrinsic phase speed $\hat{c} = c - U$ is zero. The above formulas remain valid after substituting for U by $U - c$ everywhere. As the sign of p is the sign of $\hat{c}$, it is easy to see from (3.46) that dissipating gravity waves will always accelerate the mean flow towards c. In other words, the mean-flow acceleration is always such that $\hat{c}$ is diminished. In the mountain wave case this means that the mean flow is driven towards $U = 0$, regardless of what the sign of $U(0)$ is. So wave dissipation drives the mean flow towards decreasing $\hat{c}$, which ultimately may lead to the occurrence of critical layers where $\hat{c} = 0$, and hence to increased wave dissipation, as we have seen. This is an important positive feedback cycle.

Finally, we note that wave reflection in the vertical is possible at altitudes where U reaches the upper speed limit N/k. As in the Rossby-wave case, there is no breakdown of linear theory there, although JWKB becomes invalid because $m \to 0$ at the reflection altitude.

3.9   Mean-flow feedback

Let us consider the mean-flow acceleration equation (3.54), but now written down for a wave with phase speed c:

$$\overline{u}_t(z,t) = \frac{\alpha N}{k(U(z) - c)^2} \, \overline{u'w'}(0) \exp\left(-\frac{\alpha N}{k} \int_0^z \frac{\mathrm{d}\bar{z}}{(U(\bar{z}) - c)^2} \right). \qquad (3.58)$$

From this we see that $\overline{u} - U = O(a^2 t)$, i.e. the mean-flow change grows linearly in time. For times $t = O(1)$ the mean-flow change is $O(a^2)$ as assumed in our small-amplitude scaling. However, for longer times $t = O(a^{-2})$ we can expect that the mean-flow change becomes $O(1)$, which is comparable to U. Of course, this defeats our scaling assumptions and our regular perturbation theory is not valid over such long, amplitude-dependent time scales. In principle, this requires the use of a singular perturbation theory capable of dealing with multiple time scales.

It is plausible (though difficult to prove rigorously) that the correct generalization of (3.58) valid for such long times is obtained by replacing U by $\overline{u}$ everywhere on the right-hand side. In essence, this allows the $O(1)$ background flow to evolve slowly in response to the $O(a^2)$ wave-induced forcing terms. This background evolution then feeds back into the structure of the waves themselves, and this produces an important nonlinear feedback cycle (e.g. Plumb (1977)).[4]

---

[4] The difficulty for rigorous theory is to ensure that the evolving background flow remains slowly varying relative to the waves.

Following this idea the resulting dynamical equation for $\overline{u}(z,t)$ is then

$$\overline{u}_t(z,t) - \nu\overline{u}_{zz} = -(\overline{u'w'})_z = \frac{\alpha N}{k(\overline{u}-c)^2}\,\overline{u'w'}(0)\,\exp\left(-\frac{\alpha N}{k}\int_0^z \frac{d\bar{z}}{(\overline{u}(\bar{z},t)-c)^2}\right),$$

(3.59)

which also includes vertical diffusion with constant diffusivity $\nu > 0$. This is a non-local equation for $\overline{u}$, meaning that $\overline{u}_t$ at some z depends on the values of $\overline{u}$ at all altitudes below z.

What do solutions of (3.59) look like, say for $c > 0$? It can be shown that if initially $\overline{u}(z,0) = 0$ and if the lower boundary condition $\overline{u}(0,t) = 0$, then the time evolution of (3.59) will lead towards a stable steady state such that

$$\nu\overline{u}_z = \overline{u'w'}(z) = \overline{u'w'}(0)\,\exp\left(-\frac{\alpha N}{k}\int_0^z \frac{d\bar{z}}{(\overline{u}(\bar{z},t)-c)^2}\right).$$

(3.60)

This means that $\overline{u} = c$ everywhere apart from inside a thin boundary layer near the ground. The shear at the ground $\overline{u}_z(0) = \overline{u'w'}(0)/\nu$, and hence the depth of the boundary layer is approximately $c\nu/\overline{u'w'}(0)$. Thus a single wave with $c > 0$ leads to a stable mean-flow profile with $\overline{u} = c$ outside a viscous boundary layer. This was to expected, based on the understanding that dissipating waves always drag the mean flow towards the phase speed of the wave. When we consider the mean-flow response to several waves then interesting mean-flow oscillations can occur.

3.10 Several waves and the quasi-biennial oscillation

Linear waves can be freely superimposed at $O(a)$ but we must always be careful when we compute the $O(a^2)$ momentum flux $\overline{u'w'}$, which may be affected by correlations between wave modes. Specifically, consider two waves with parameters $\{k_1, c_1\}$ and $\{k_2, c_2\}$, respectively. If $|k_1| \neq |k_2|$ then the wave modes will be orthogonal with respect to x-averaging and hence

$$\overline{u'w'} = \overline{u_1'w_1'} + \overline{u_2'w_2'}.$$

(3.61)

On the other hand, if the wavenumber magnitudes are equal then $\overline{u'w'}$ will oscillate in time around a mean value given by (3.61) and the oscillations will have frequencies $|k(c_1 \pm c_2)|$. In practice one assumes that only the time-averaged value of $\overline{u'w'}$ matters for the long-time mean-flow forcing and hence (3.61) is assumed to hold.

Consider now the specific case where $k_1 = k_2 = k > 0$ and $c_1 = -c_2 = c > 0$ and the wave amplitudes are equal such that $\overline{u_1'w_1'}(0) = A > 0$ and $\overline{u_2'w_2'}(0) = -A$. This corresponds to a *standing wave* pattern $\psi' \propto \cos(kx)\cos(ckt)$ at fixed z. The mean-flow equation is given by (3.59) written down for forcing due to two waves, i.e.

$$\overline{u}_t - \nu\overline{u}_{zz} = \frac{\alpha N}{k}\left\{\frac{A}{(\overline{u}-c)^2}\exp\left(-\frac{\alpha N}{k}\int_0^z \frac{d\bar{z}}{(\overline{u}-c)^2}\right) - \frac{A}{(\overline{u}+c)^2}\exp\left(-\frac{\alpha N}{k}\int_0^z \frac{d\bar{z}}{(\overline{u}+c)^2}\right)\right\}.$$

(3.62)

This complicated-looking equation can be easily integrated numerically. Clearly, $\overline{u} = 0$ is a trivial equilibrium state as both waves then neutralize each other at all z, meaning that their respective dissipation exerts equal and opposite forces onto the mean flow.

However, it can be seen that this is not a stable equilibrium configuration. Indeed, if there is a small positive disturbance to $\bar{u}$ at some location then the first wave will be preferentially dissipated there, because $\bar{u} - c$ will have been reduced there. This means that $\bar{u}_t > 0$ there, amplifying the original disturbance and hence showing that the equilibrium is unstable. Furthermore, because the first wave dissipates stronger at this location, the second wave will dissipate stronger than the first wave in the region *above* the original positive disturbance. This means that the region above will experience an acceleration $\bar{u}_t < 0$. A modicum of further analysis reveals that the zero-wind line $\bar{u} = 0$ above the original disturbance travels downward in time as do the positive and negative mean-flow regions themselves.

In summary, the zero-flow equilibrium is unstable and spontaneously breaks down into an oscillatory flow pattern that travels downward. The frequency of these mean-flow oscillations depends on the details of the incident waves and is inversely proportional to $O(a^2)$, i.e. stronger waves lead to more rapid oscillations.

This spontaneous occurrence of a mean-flow oscillation is believed to be fundamental to the so-called "quasi-biennial oscillation" in the lower stratosphere (Baldwin et al. (2001)). This is a periodic reversal of the zonal mean winds between about 18-30 km in the equatorial stratosphere. The period of these carefully observed oscillations is about 26-27 months, which is *not* a sub-harmonic of the annual cycle. The zonal mean wind pattern is observed to travel downwards at a speed of about 1km per month. Extensions of the above theory for more than two waves and for more than one kind of wave (i.e. including Rossby waves) are believed to offer the best scientific explanation for these peculiar wind oscillations.

4 Circulation and pseudomomentum

Small-amplitude equations such as (3.28) can turn out to be leading-order versions of fully nonlinear conservation laws, i.e. laws that can in principle be formulated without restriction to small wave amplitude a. This points towards significant extensions of our small-amplitude theory. We will not pursue these extension here, but we will illustrate one way in which (3.28) can be shown to be a small-amplitude version of Kelvin's circulation theorem (e.g. Bretherton (1969)). This will provide a satisfying physical interpretation of the pseudomomentum p and opens a perspective towards extensions of wave–mean interaction theory beyond simple geometry. Such extensions are the subject of on-going research (e.g. Bühler and McIntyre (2004), Bühler and McIntyre (2003), Bühler (2000), McIntyre and Norton (1990)).

4.1 Kelvin's circulation theorem

Consider a closed oriented circuit C lying in the fluid. The circulation Γ around this circuit is defined by the closed line integral

$$\Gamma = \oint_C \boldsymbol{u} \cdot \mathrm{d}\boldsymbol{x}. \tag{4.1}$$

If C is a material circuit (i.e. it moves with the flow) then the material time derivative of $d\boldsymbol{x}$ is $d\boldsymbol{u}$ and that of Γ can then be shown to be

$$\frac{d\Gamma}{dt} = \oint_C \frac{D\boldsymbol{u}}{Dt} \cdot d\boldsymbol{x} + \oint_C \boldsymbol{u} \cdot d\boldsymbol{u} = -\oint_C \frac{\nabla p}{\rho} \cdot d\boldsymbol{x} + 0, \qquad (4.2)$$

where we neglect background rotation at the moment but allow for variable density ρ.

If the integrand is a perfect differential then the closed line integral vanishes and Γ does not change. This can happen in three physically interesting ways.

- First, ρ could be a global constant, as in homogeneous incompressible flow.
- Second, ρ could be a global function of pressure, as in so-called barotropic flow.
- Third, ρ could be general function of pressure and a second thermodynamic variable such as specific entropy. Then, *if C lies inside a surface of constant entropy* the integrand is a perfect differential on C and hence the integral vanishes. Furthermore, if the specific entropy is materially invariant then C will continue to lie within a surface of constant entropy, and hence $\Gamma = \text{const.}$ holds for circuits lying within entropy surfaces. This is the case of most interest in atmosphere and ocean fluid dynamics.

In all three cases we have the celebrated *Kelvin circulation theorem*

$$\frac{d\Gamma}{dt} = 0. \qquad (4.3)$$

If there is background rotation with Coriolis vector $\boldsymbol{f}$ then the above considerations remain trivially true provided that $\boldsymbol{u}$ is replaced in (4.1) by the absolute velocity $\boldsymbol{u} + 0.5\boldsymbol{f} \times \boldsymbol{x}$. With this simple extension in mind we will continue to neglect background rotation.

In the Boussinesq system the stratification $\theta = b + N^2 z$ plays the same role as the entropy above. This can be shown directly from the Boussinesq equations, which yield

$$\frac{d\Gamma}{dt} = -\oint_C \frac{\nabla p}{\rho_0} \cdot d\boldsymbol{x} + \oint_C b\hat{\boldsymbol{z}} \cdot d\boldsymbol{x} = 0 + \oint_C b\,dz \qquad (4.4)$$

The remaining integral can be re-written as

$$\oint_C (b + N^2 z)dz - \oint_C N^2 z\,dz. \qquad (4.5)$$

The second integral is clearly zero and if θ is constant on C then it can be pulled out of the first integral, which is then zero as well. Therefore circulation is indeed conserved on circuits lying within stratification surfaces. In a periodic domain a line traversing the the length of the domain can be viewed as a closed circuit and hence in the two-dimensional Boussinesq equations circulation is conserved on the kind of undulating material lines that were considered before.

Clearly, to make use of the circulation theorem we must have a way to follow material displacements induced by the waves. In linear wave theory this is achieved by defining the three-dimensional linear particle displacement vector

$$\boldsymbol{\xi}' = (\xi', \eta', \zeta') \quad \text{via} \quad D_t \boldsymbol{\xi}' = \boldsymbol{u}' \qquad (4.6)$$

in terms of the linear wave velocities u'. This defines ξ' up to an integration constant that has to be determined from other considerations. Often, the flow can be imagined to have started from rest and then the integration constant is zero. For a plane wave the simple relation $-i\hat{\omega}\xi' = u'$ holds. A particle that was initially at rest at location x will later be displaced by the waves to a position $x + \xi'$. Advected fields give access to some of the components of ξ'. For instance, in the Boussinesq equations we have $\zeta' = -b'/N^2$.

Let us now denote by C^ξ a line of constant stratification θ, which is a material contour as well. In the undisturbed configuration (i.e. without waves) this contour is a flat line $z = \text{const.}$, which we will denote by C. Kelvin's circulation theorem then tells us that the circulation along C^ξ is exactly constant. To $O(a)$ the displaced contour C^ξ is given by the "lifting" map

$$x \quad \rightarrow \quad x + \xi'(x, t) \tag{4.7}$$
$$\Leftrightarrow \quad (x, z) \quad \rightarrow \quad (x + \xi'(x, z, t), \, z + \zeta'(x, z, t)) \tag{4.8}$$

based on the $O(a)$ linear displacements ξ'. Here z is held constant, x varies from 0 to L, and the induced map "lifts" positions from the undisturbed contour C to the displaced contour C^ξ. The key step is now to express the conserved circulation

$$\Gamma = \oint_{C^\xi} u\,\mathrm{d}x + w\,\mathrm{d}z \tag{4.9}$$

in terms of an integral over the undisturbed contour C. The lifting map (4.7) reduces this to a simple problem of variable transformation in the line integral:

$$\Gamma = \oint_C u^\xi (\mathrm{d}x)^\xi + w^\xi (\mathrm{d}z)^\xi, \tag{4.10}$$

where we have used the nifty shorthand $u^\xi(x, t) \equiv u(x + \xi', t)$. Here the line element transforms as

$$(\mathrm{d}x)^\xi = \mathrm{d}x + \mathrm{d}\xi' \quad = \quad \mathrm{d}x(1 + \xi'_x) \tag{4.11}$$
$$(\mathrm{d}z)^\xi = \mathrm{d}z + \mathrm{d}\zeta' \quad = \quad \mathrm{d}x\,\zeta'_x \tag{4.12}$$

because $\mathrm{d}z = 0$ on the flat contour C. The integral over C is an integral over x at fixed z, i.e.

$$\Gamma = \int_0^L \left[u^\xi(1 + \xi'_x) + w^\xi\,\zeta'_x \right] \mathrm{d}x \quad = \quad L\,\overline{\left[u^\xi(1 + \xi'_x) + w^\xi\,\zeta'_x \right]} \tag{4.13}$$
$$= \quad L\overline{u}^L + L\,\overline{\left[u^\xi\xi'_x + w^\xi\,\zeta'_x \right]} \tag{4.14}$$

where we have introduced the *Lagrangian-mean* velocity $\overline{u}^L \equiv \overline{u^\xi}$, which is a particle-following average velocity.[5] To compute Γ correct to $O(a^2)$ requires only $O(a)$ accuracy

---

[5]In general, $\overline{u}^L$ and the Eulerian-mean $\overline{u}$ differ by an $O(a^2)$ Stokes drift correction. However, the Stokes drift can be computed from the linear wave solution, so knowledge of $\overline{u}$ implies knowledge of $\overline{u}^L$ and vice versa. The decision of which mean velocity to use in a given problem can therefore be based on convenience.

in the remaining $\boldsymbol{u}^\xi$ terms, because $\boldsymbol{\xi}' = O(a)$. Also, $\boldsymbol{u}^\xi = \boldsymbol{u}' + O(a^2)$ if the background flow is constant and hence we obtain that

$$\Gamma = L\bar{u}^L + L\,\overline{[u'\xi_x' + w'\zeta_x']} \equiv L\left(\bar{u}^L - \mathsf{p}\right) \tag{4.15}$$

provided we *define* the pseudomomentum to be

$$\boxed{\mathsf{p} \equiv -\overline{\xi_x'u'} - \overline{\zeta_x'w'}}. \tag{4.16}$$

This new Lagrangian definition is consistent with the earlier Eulerian pseudomomentum definition (3.25), i.e.

$$-\overline{\zeta'(u_z' - w_x')} = -(\overline{\zeta'u'})_z + \overline{\zeta_z'u'} + \overline{\zeta'w_x'} = -(\overline{\zeta'u'})_z - \overline{\xi_x'u'} - \overline{\zeta_x'w'}. \tag{4.17}$$

The extra term $(\overline{\zeta'u'})_z$ vanishes for plane waves and due to its flux divergence form it does not upset the conservation law for p. Now let us consider what (4.15) implies. Because the circulation Γ is constant we must have

$$\bar{u}_t^L - \mathsf{p}_t = 0, \tag{4.18}$$

which is our usual mean-flow equation. This now arises naturally from the finite-amplitude conservation law for circulation. This is a different physical concept than momentum: pseudomomentum is closer related to circulation than to momentum.

4.2   Vectorial pseudomomentum

The pseudomomentum definition (4.16) suggests a natural extension to a vectorial pseudomomentum with components

$$\mathsf{p}_i \equiv -\overline{(\xi_{j,i}'u_j')} \quad \text{(summation understood)} \tag{4.19}$$

such that

$$\Gamma = \oint_{C^\xi} \boldsymbol{u} \cdot \mathrm{d}\boldsymbol{x} = \oint_C \left(\bar{\boldsymbol{u}}^L - \boldsymbol{\mathsf{p}}\right) \cdot \mathrm{d}\boldsymbol{x} \tag{4.20}$$

holds by construction for circuits C that move with $\bar{\boldsymbol{u}}^L$. This leads to useful extensions of the classical wave–mean interaction theory to situations with non-simple geometry. One example of this is the generation of mean-flow vortices by breaking ocean waves on a beach (Bühler and Jacobson (2001)).

In general, what emerges from (4.19-4.20) is that pseudomomentum chose its name wisely: it is closely linked to fluid circulation and not to momentum. It measures the wave contribution to the circulation along stratification contours. The remaining contribution to the circulation is taken up by the Lagrangian-mean flow $\bar{\boldsymbol{u}}^L$.

5 Wave-driven global atmospheric circulations

Dissipating Rossby and gravity waves are essential contributors to the global zonal-mean circulation in the middle atmosphere, between 10-100 km or so (e.g. Andrews et al.

(1987), McIntyre (2000)). This region consists of the stratosphere between 10-60 km and the mesosphere above, between 60-100 km. Rossby waves dominate in the stratosphere and gravity waves in the mesosphere. The lower atmosphere below 10 km is called the troposphere, where our weather lives and where most of the atmospheric mass resides. However, the motion of the middle atmosphere is crucial for long-term trends in climate, such as the motion of long-lived chemicals responsible for the ozone hole. Now, a key to understanding the wave-induced circulation mechanisms is to note that on a rotating planet a mid-latitude zonal-mean westward (or retrograde) force reduces the absolute angular momentum of a ring of particles (i.e. a set of particles that initially shared the same altitude and latitude) and this drives that ring of particles poleward, i.e. towards the rotation axis. The mass flux of this poleward motion is compensated by sinking motion over the pole and rising motion over the equator. This works on both hemispheres and is sometimes called "gyroscopic pumping". Conversely, a zonal-mean eastward (or prograde) force increases the angular momentum and hence drives the ring of particles equatorward.

One can show that three-dimensional Rossby waves behave in fundamentally the same way as the two-dimensional waves we have investigated. In particular, dissipating Rossby waves always exert a retrograde force, and Rossby waves can be generated by flow over an undulating boundary, usually over mountains below. There is persistent Rossby-wave-induced retrograde forcing in the stratosphere in both hemispheres. This leads to poleward stratospheric motion at about 30 km, rising motion between 10-30 km over the equator, and sinking motion over the poles. This is called the stratospheric *Brewer-Dobson* circulation. Despite appearance, this circulation can not be explained by hot air rising over the equator and drifting polewards. This is because one has to explain the angular momentum change that allows particles to drift poleward on a rotating planet. This *requires* wave-induced forces to provide the necessary torque.

In the stratosphere Rossby waves are the most important waves, but higher up, in the mesosphere, the gravity waves dominate over Rossby waves. The density throughout the middle atmosphere decays roughly as

$$\rho(z) = \rho_0 \exp(-z/H_S) \tag{5.1}$$

where the density scale height $H_S \approx 7$km. It can be shown that this implies that Boussinesq gravity waves are modified such that a steady non-dissipating wave now obeys $\rho(z)\overline{u'w'} = $ const. This implies that the particle velocities u' and w' *increase* with altitude. This leads to very large wave amplitudes at very high altitudes, especially in the so-called mesosphere, which begins around 60km or so. The large wave amplitudes lead to gravity-wave breaking in the mesosphere. Furthermore, the kinematic viscosity $\nu = \eta/\rho(z)$, where η is the dynamic viscosity, which does not vary much. This means that ν becomes very large and waves are subject to very strong diffusion. In practice, it is assumed that gravity waves must break down in the mesosphere due to a combination of nonlinear turbulence and viscous dissipation.

Mountain gravity waves that reach the mesosphere hence dissipate and exert a force on the mean flow that drives $\overline{u}$ to zero. This "gravity-wave drag" is known to be crucial for the observed structure of the zonal wind: without this drag there would be enormously large winds at these altitudes, which are not observed.

Now, there are increasing stratospheric temperatures from winter to summer hemisphere, meaning that there is a robust latitudinal gradient of zonal-mean temperature. On a rotating planet this is concomitant with a vertical shear of the zonal wind $\bar{u}$ (this link is described by the "thermal wind relation"). It turns out that this leads to strong prograde wind $\bar{u} > 0$ in the winter stratosphere and strong retrograde wind $\bar{u} < 0$ in the summer stratosphere. Gravity-wave drag that drives the wind to zero is hence retrograde in the winter hemisphere and prograde in the summer hemisphere! There is a second mechanism, which is not tied to zero-phase speed mountain waves but which leads to the same conclusion: critical-layer filtering in the stratosphere. The prograde winds in the winter stratosphere preferentially filter gravity waves with prograde phase velocities $c > 0$ and vice versa in the summer stratosphere. That results in the preferential transmission of retrograde gravity waves into the winter mesosphere and of prograde gravity waves into the summer mesosphere. The dissipation of these waves then gives the same result: retrograde drag in winter and prograde drag in summer.

The net result is a poleward flow in the winter mesosphere and an equatorward flow in the summer mesosphere. Together, this gives a net flow from summer to winter mesosphere, above about 60 km. There is no rising motion over the equator now, but there is sinking motion over the winter pole and rising motion over the summer pole to close the mass flux budget. This is called the mesospheric *Murgatroyd–Singleton* circulation. The rising motion over the summer pole is particularly important, because the adiabatic expansion of the rising air produces the *coldest* temperatures on Earth in the summer polar mesosphere even though this is the *sunniest* place on Earth as well. Temperatures as low $-163^{\circ}C$ have been recorded and the extreme cold gives rise to "noctilucent clouds", which glow in electric blue at around 85 km and are made of ice crystals.

Finally, we note that gravity waves are far too small in scale (especially vertical scale) to be directly resolved in numerical models and hence gravity-wave drag must be put in by hand, based on the kind of theory we are studying here. This is an area of active research (Fritts and Alexander (2003), Kim et al. (2003)).

Bibliography

D. G. Andrews, J. R. Holton, and C. B. Leovy. *Middle Atmosphere Dynamics*. Academic Press, 1987.

M. P. Baldwin, L. J. Gray, T. J. Dunkerton, K. Hamilton, P. H. Haynes, W. J. Randel, J. R. Holton, M. J. Alexander, I. Hirota, T. Horinouchi, D. B. A. Jones, J. S. Kinnersley, C. Marquardt, K. Sato, and M. Takahashi. The quasi-biennial oscillation. *Revs. Geophys.*, 39:179–229, 2001.

F. P. Bretherton. On the mean motion induced by internal gravity waves. *J. Fluid Mech.*, 36:785–803, 1969.

O. Bühler. On the vorticity transport due to dissipating or breaking waves in shallow-water flow. *J. Fluid Mech.*, 407:235–263, 2000.

O. Bühler and T. E. Jacobson. Wave-driven currents and vortex dynamics on barred beaches. *J. Fluid Mech.*, 449:313–339, 2001.

O. Bühler and M. E. McIntyre. Remote recoil: a new wave-mean interaction effect. *J. Fluid Mech.*, 492:207–230, 2003.

O. Bühler and M. E. McIntyre. Wave capture and wave–vortex duality. Submitted to *J. Fluid Mech.* Preprint available at http://www.cims.nyu.edu/~obuhler., 2004.

P. G. Drazin and W. H. Reid. *Hydrodynamic Stability*. Cambridge University Press, 1981.

D. C. Fritts and M. J. Alexander. Gravity-wave dynamics and effects in the middle atmosphere. *Revs. Geophys.*, 41(1), doi:10.1029/2001RG000106., 2003.

P. H. Haynes. Nonlinear instability of a rossby-wave critical layer. *J. Fluid Mech.*, 161: 493–511, 1985.

P. H. Haynes. On the instability of sheared disturbances. *J. Fluid Mech.*, 175:463–478, 1987.

P. H. Haynes. Critical layers. In J. R. Holton, J. A. Pyle, and J. A. Curry, editors, *Encyclopedia of Atmospheric Sciences*. London, Academic/Elsevier, 2003.

I Held. The general circulation of the atmosphere. *GFD lecture notes, ed. Rick Salmon. http://gfd.whoi.edu/proceedings/2000/PDF/lectures2000.pdf*, 2000.

C. O. Hines. Earlier days of gravity waves revisited. *Pure and Applied Geophysics*, 130: 151–170, 1989.

P. D. Killworth and M. E. McIntyre. Do rossby-wave critical layers absorb, reflect or over-reflect? j. fluid mech. *449–492.*, , 1985.

Y.-J. Kim, S. D. Eckermann, and H.-Y. Chun. An overview of the past, present and future of gravity-wave drag parametrization for numerical climate and weather prediction models. *Atmos. –Oc.*, 41:65–98, 2003.

M. J. Lighthill. *Waves in Fluids*. Cambridge University Press, Cambridge, 1978.

M. E. McIntyre. On global-scale atmospheric circulations. In G. K. Batchelor, H. K. Moffatt, and M. G. Worster, editors, *Perspectives in Fluid Dynamics: A Collective Introduction to Current Research*, pages 557–624. Cambridge, University Press, 631 pp., 2000.

M. E. McIntyre and W. A. Norton. Dissipative wave-mean interactions and the transport of vorticity or potential vorticity. *J. Fluid Mech.*, 212, 403–435. Corrigendum 220, 693., 1990.

J. Pedlosky. *Geophysical Fluid Dynamics (2nd edition)*. Springer-Verlag, New York, 1987.

R. A. Plumb. The interaction of two internal waves with the mean flow: implications for the theory of the quasi-biennial oscillation. *J. Atmos. Sci.*, 34:1847–1858, 1977.

C. Staquet and J. Sommeria. Internal gravity waves: From instabilities to turbulence. *Ann. Rev. Fluid Mech.*, 34:559–594, 2002.

Wave Turbulence with Applications to Atmospheric and Oceanic Waves

V. Zeitlin

LMD-ENS, Paris, France

Abstract We give a review of the main ideas and methods of the wave-turbulence theory and their applications to geophysical fluid dynamics. After having introduced the basic hypotheses leading to kinetic equations for ensembles of weakly nonlinear waves we explain the methods of finding stationary solutions, both for isotropic and weakly anisotropic dispersion relations. We then show how the method can be applied to waves in the atmosphere and ocean and review the known results in this area. Turbulence of the short inertia-gravity waves in the rotating shallow water model is considered at tempered latitudes and in the tropical region, and corresponding stationary spectra are found. Turbulence of the Rossby waves in the same model is also reviewed. Finally, the problem of turbulence of weakly nonlinear internal gravity waves in the continuously stratified fluid is addressed.

1 Basic ideas and methods of wave (weak) turbulence: surface waves

1.1   The main hypotheses and ideas of the wave turbulence approach

Consider *ensemble* of *large number* of *weakly nonlinear* (i.e. small-amplitude) *harmonic* waves taken in the complex form:

$$a_{\mathbf{k}}^{(0)} e^{i(\mathbf{k}\cdot\mathbf{x} - \omega(\mathbf{k})t)}. \tag{1.1}$$

The waves are solutions of the *linearized* dynamical equations describing some non-dissipative medium which occupies the whole space (no boundaries). The properties of the medium determine the *dispersion relation*: dependence of wave-frequencies ω on wave-numbers $\mathbf{k}$: $\omega = \omega(\mathbf{k})$. This equation provides a *spectrum* of infinithesimal excitations of the medium. The equations (the model) of the medium are, generally, *nonlinear* and nonlinear terms engender interactions of harmonic waves. The nonlinearities of the system may be quadratic, cubic, etc in wave- amplitudes. In the *weak turbulence* approach the wave-amplitudes $a_{\mathbf{k}}$ are always supposed to be small, and hence only lower-order nonlinearities are taken into account. Nonlinear effects are considered as *small perturbations* to the linear wave field. The supposed weakness of wave interactions is essential.

If the dispersion relation (spectrum) is of *decay* type, i.e. the following equation:

$$\omega(\mathbf{k}_1 + \mathbf{k}_2) = \omega(\mathbf{k}_1) + \omega(\mathbf{k}_2) \tag{1.2}$$

has (non-zero) solutions, then it is sufficient to take into account only the lowest-order interactions. Otherwise (spectrum of *non-decay* type), the next-order interactions should be retained.

A systematic way to realize the weak turbulence approach is through Hamiltonian description of the medium. If all physical fields are Fourier-transformed in space, the linear equations for their Fourier-transforms $a_\mathbf{k}$ giving harmonic waves (1.1) are:

$$\dot{a}_\mathbf{k} + i\omega(\mathbf{k})a_\mathbf{k} = 0; \quad a_\mathbf{k} = a_\mathbf{k}^{(0)} e^{-i\omega(\mathbf{k})t}. \tag{1.3}$$

Here and below dot denotes time-derivative. These equations are Hamiltonian:

$$\dot{a}_\mathbf{k} = -i\frac{\delta H_0}{\delta a_\mathbf{k}^*}, \tag{1.4}$$

with

$$H_0 = \int d\mathbf{k}\,\omega(\mathbf{k})\, a_\mathbf{k} a_\mathbf{k}^*. \tag{1.5}$$

The Hamiltonian of the free waves H_0 is the energy of an ensemble of non-interacting harmonic oscillators with frequencies $\omega(\mathbf{k})$.

We limit ourselves in this section by a pair of real functions describing the medium and, hence, by a single complex amplitude $a_\mathbf{k}$. For instance, for surface waves in the fluid the two real functions are the free surface displacement and the value of the velocity potential at the surface. If more fields are necessary to describe the medium, several complex Fourier-amplitudes should be introduced, with obvious modifications below.

The full Hamiltonian contains an interaction term: $H = H_0 + H_{int}$ with H_{int} which may be expanded in powers of $a_\mathbf{k}$ and its complex conjugate $a_\mathbf{k}^*$: $H_{int} = H_3 + H_4 + ...$ with (for stable homogeneous medium):

$$H_3 = \frac{1}{2}\int d\mathbf{k}d\mathbf{k}_1 d\mathbf{k}_2\, \delta(\mathbf{k} - \mathbf{k}_1 - \mathbf{k}_2)\, V_{\mathbf{k}\mathbf{k}_1\mathbf{k}_2} a_\mathbf{k}^* a_{\mathbf{k}_1} a_{\mathbf{k}_2} + c.c., \tag{1.6}$$

$$H_4 = \frac{1}{2}\int d\mathbf{k}d\mathbf{k}_1 d\mathbf{k}_2 d\mathbf{k}_3\, \delta(\mathbf{k} + \mathbf{k}_1 - \mathbf{k}_2 - \mathbf{k}_3)\, W_{\mathbf{k}\mathbf{k}_1\mathbf{k}_2\mathbf{k}_3} a_\mathbf{k}^* a_{\mathbf{k}_1}^* a_{\mathbf{k}_2} a_{\mathbf{k}_3}. \tag{1.7}$$

The full equations of motion are, correspondingly,

$$\dot{a}_\mathbf{k} = -i\omega(\mathbf{k})a_\mathbf{k} - i\frac{\delta H_{int}}{\delta a_\mathbf{k}^*}. \tag{1.8}$$

Remember that amplitudes are always supposed to be small and, hence, the interaction Hamiltonian gives only small corrections to the linear solutions.

The main hypothesis of the weak turbulence is that weak interactions of a large number of waves lead to *phase randomization* (a central limit theorem-like argument) and *Gaussian statistics* of the wave field. As usual in statistical approach, *ergodic hypothesis* is supposed to be valid. From the full dynamical description (1.8) one passes then to the *statistical* description, where the system is described by a set of correlation functions of complex amplitudes. Ensemble averaging will be denoted from now on by $\langle ... \rangle$.

Gaussianity for spatially uniform medium means that all odd-order correlators vanish and that all even-order correlators are expressed in terms of the (real) quadratic one:

$$\langle a_{\mathbf{k}} a_{\mathbf{k'}} \rangle = 0, \quad \langle a_{\mathbf{k}} a_{\mathbf{k'}}^* \rangle = N(\mathbf{k})\delta(\mathbf{k} - \mathbf{k'}),$$
$$\langle a_{\mathbf{k}_1}^* a_{\mathbf{k}_2}^* a_{\mathbf{k}_3} a_{\mathbf{k}_4} \rangle = N(\mathbf{k}_1)N(\mathbf{k}_2)\left(\delta(\mathbf{k}_1 - \mathbf{k}_3)\delta(\mathbf{k}_2 - \mathbf{k}_4) + \delta(\mathbf{k}_1 - \mathbf{k}_4)\delta(\mathbf{k}_2 - \mathbf{k}_3)\right).$$
$$(1.9)$$

As all correlators are expressed in terms of the quadratic one, i. e. in terms of $N(\mathbf{k})$, the main goal of the statistical theory is to determine the evolution of this quantity which is given by *kinetic equation*:

$$\dot{N}(\mathbf{k}) = \mathcal{I}\left[N(\mathbf{k})\right], \qquad (1.10)$$

where expression in the r.h.s. is called *collision integral*.

1.2   Kinetic equations for decay and non-decay dispersion laws

A quantum-mechanical analogy allows to establish heuristically such equation both for decay and non-decay spectra (for direct derivation see Section 4). Indeed, $N(\mathbf{k})$ is the distribution function for wave amplitudes in the phase-space, and may be interpreted as density of some quasi-particles in the phase-space. The changes of this density are due to the difference between inflow and outflow of quasi-particles in any element of the phase-space. The complex wave amplitudes $a_{\mathbf{k}}$, $a_{\mathbf{k}}^*$ are interpreted as annihilation and creation operators of quasi-particles, respectively.

The quantum-mechanical probability of transition between two given states $(i \rightarrow f)$ is proportional to the matrix element of the interaction Hamiltonian between these states times delta-function corresponding to energy and other integrals of motion:

$$W = \frac{2\pi}{\hbar}\left|(f|H_{int}|i)\right|^2 \delta(E_f - E_i). \qquad (1.11)$$

(The Planck constant may be absorbed in what follows by proper renormalization of wave-amplitudes and frequencies.) The matrix elements of creation and annihillation operators are:

$$(N-1|a|N) = \sqrt{N}; \quad (N+1|a^*|N) = \sqrt{N+1}. \qquad (1.12)$$

Take first a **decay spectrum case**, i.e. a cubic interaction Hamiltonian. Then the term $V_{\mathbf{k}\mathbf{k}_1\mathbf{k}_2} a_{\mathbf{k}}^* a_{\mathbf{k}_1} a_{\mathbf{k}_2}$ corresponds to the process of annihilation of two quasi-particles with momenta $\mathbf{k}_1$, $\mathbf{k}_2$, and creation of a quasi-particle with momentum $\mathbf{k}$. Hence, this term will give an inflow probability $2\pi |V_{\mathbf{k}\mathbf{k}_1\mathbf{k}_2}|^2 (N(\mathbf{k})+1)N(\mathbf{k}_1)N(\mathbf{k}_2)$. Its complex conjugate in H_3 (cf. (1.6)) will give an outflow term with a factor $N(\mathbf{k})(N(\mathbf{k}_1)+1)(N(\mathbf{k}_2)+1)$ with the same conservation laws and the same factor $|V|^2$. Hence, the occupation numbers N enter the quantum collision integral in the following combination:

$$f_{\mathbf{k}\mathbf{k}_1\mathbf{k}_2}^{QM} = (N(\mathbf{k}) + 1)N(\mathbf{k}_1)N(\mathbf{k}_2) - N(\mathbf{k})(N(\mathbf{k}_1) + 1)(N(\mathbf{k}_2) + 1) \qquad (1.13)$$

The classical limit corresponds to $N \gg 1$, and the classical analog of (1.13) is:

$$f_{\mathbf{k}\mathbf{k}_1\mathbf{k}_2} = N(\mathbf{k}_1)N(\mathbf{k}_2) - N(\mathbf{k})\left(N(\mathbf{k}_1) + N(\mathbf{k}_2)\right). \qquad (1.14)$$

To get the full collision integral, the terms obtained by cyclic permutation of wave-vectors, with corresponding changes of sign as inflow and outflow change roles, should be added and we get:

$$\mathcal{I}^{(3)}\left[N(\mathbf{k})\right] = \int d\mathbf{k}_1 d\mathbf{k}_2 \left[W_{\mathbf{k}\mathbf{k}_1\mathbf{k}_2} f_{\mathbf{k}\mathbf{k}_1\mathbf{k}_2} - W_{\mathbf{k}_1\mathbf{k}_2\mathbf{k}} f_{\mathbf{k}_1\mathbf{k}_2\mathbf{k}} - W_{\mathbf{k}_2\mathbf{k}\mathbf{k}_1} f_{\mathbf{k}_2\mathbf{k}\mathbf{k}_1}\right], \qquad (1.15)$$

where

$$W_{\mathbf{k}\mathbf{k}_1\mathbf{k}_2} = 2\pi \left|V_{\mathbf{k}\mathbf{k}_1\mathbf{k}_2}\right|^2 \delta(\mathbf{k} - \mathbf{k}_1 - \mathbf{k}_2)\delta(\omega(\mathbf{k}) - \omega(\mathbf{k}_1) - \omega(\mathbf{k}_2)). \qquad (1.16)$$

For the **non-decay spectrum case**, the interaction Hamiltonian, in general, contains both H_3 and H_4 contributions. The cubic terms in the Hamiltonian may be removed by the following canonical transformation:

$$\begin{aligned}
a_{\mathbf{k}} \quad \rightarrow \quad & a_{\mathbf{k}} - \int d\mathbf{k}_1 d\mathbf{k}_2 \, \delta(\mathbf{k} - \mathbf{k}_1 - \mathbf{k}_2)\frac{V_{\mathbf{k}\mathbf{k}_1\mathbf{k}_2}}{\omega(\mathbf{k}) - \omega(\mathbf{k}_1) - \omega(\mathbf{k}_2)} a_{\mathbf{k}_1} a_{\mathbf{k}_2} \\
& -2 \int d\mathbf{k}_1 d\mathbf{k}_2 \, \delta(\mathbf{k} + \mathbf{k}_1 - \mathbf{k}_2)\frac{V_{\mathbf{k}\mathbf{k}_1\mathbf{k}_2}}{-\omega(\mathbf{k}) + \omega(\mathbf{k}_1) - \omega(\mathbf{k}_2)} a_{\mathbf{k}_1}^* a_{\mathbf{k}_2}.
\end{aligned}$$

$$(1.17)$$

A normal form of quartic Hamiltonian thus follows:

$$\begin{aligned}
H \quad = \quad & \int d\mathbf{k}\,\omega(\mathbf{k})\, a_{\mathbf{k}} a_{\mathbf{k}}^* \\
& + \frac{1}{2} \int d\mathbf{k} d\mathbf{k}_1 d\mathbf{k}_2 d\mathbf{k}_3 \, \delta(\mathbf{k} + \mathbf{k}_1 - \mathbf{k}_2 - \mathbf{k}_3)\, T_{\mathbf{k}\mathbf{k}_1\mathbf{k}_2\mathbf{k}_3}\, a_{\mathbf{k}}^* a_{\mathbf{k}_1}^* a_{\mathbf{k}_2} a_{\mathbf{k}_3},
\end{aligned} \qquad (1.18)$$

where

$$\begin{aligned}
T_{\mathbf{k}\mathbf{k}_1\mathbf{k}_2\mathbf{k}_3} \quad = \quad & -2\frac{V_{\mathbf{k}+\mathbf{k}_1,\mathbf{k},\mathbf{k}_1} V_{\mathbf{k}_2+\mathbf{k}_3,\mathbf{k}_2,\mathbf{k}_3}^*}{\omega(\mathbf{k}+\mathbf{k}_1) - \omega(\mathbf{k}) - \omega(\mathbf{k}_1)} - 2\frac{V_{\mathbf{k},\mathbf{k}_2,\mathbf{k}-\mathbf{k}_2} V_{\mathbf{k}_3,\mathbf{k}_1,\mathbf{k}_3-\mathbf{k}_1}^*}{\omega(\mathbf{k}_3-\mathbf{k}_1) + \omega(\mathbf{k}_1) - \omega(\mathbf{k}_3)} \\
& -2\frac{V_{\mathbf{k},\mathbf{k}_3,\mathbf{k}-\mathbf{k}_3} V_{\mathbf{k}_2,\mathbf{k}_1,\mathbf{k}_2-\mathbf{k}_1}^*}{\omega(\mathbf{k}_2-\mathbf{k}_1) + \omega(\mathbf{k}_1) - \omega(\mathbf{k}_2)} - 2\frac{V_{\mathbf{k}_1,\mathbf{k}_3,\mathbf{k}_1-\mathbf{k}_3} V_{\mathbf{k}_2,\mathbf{k},\mathbf{k}_2-\mathbf{k}}^*}{\omega(\mathbf{k}_3-\mathbf{k}_1) + \omega(\mathbf{k}_1) - \omega(\mathbf{k}_3)} \\
& -2\frac{V_{\mathbf{k}_1,\mathbf{k}_2,\mathbf{k}_1-\mathbf{k}_2} V_{\mathbf{k}_3,\mathbf{k},\mathbf{k}_3-\mathbf{k}}^*}{\omega(\mathbf{k}_2-\mathbf{k}_1) + \omega(\mathbf{k}_1) - \omega(\mathbf{k}_2)} + W_{\mathbf{k}\mathbf{k}_1\mathbf{k}_2\mathbf{k}_3}.
\end{aligned} \qquad (1.19)$$

In order to reconstruct the collision integral, the quantum analogy, again, works. Instead of $2 \rightarrow 1$ fusion and $1 \rightarrow 2$ decay processes for quasi-particles, only $2 \rightarrow 2$ processes are allowed by the Hamiltonian (1.18), either direct ones, given by the W - term in (1.19), or two-stage ones, given by VV^* - terms. By associating a factor $N + 1$ to created and a factor N to annihilated in this process particles we get a counterpart of the quantum factor (1.13):

$$f^{QM}_{\mathbf{k}\mathbf{k}_1\mathbf{k}_2\mathbf{k}_3} = (N(\mathbf{k})+1)N(\mathbf{k}_1)+1)N(\mathbf{k}_2)N(\mathbf{k}_3)-N(\mathbf{k})N(\mathbf{k}_1)(N(\mathbf{k}_2)+1)(N(\mathbf{k}_3)+1) \quad (1.20)$$

which gives a classical one at $N \gg 1$:

$$\begin{aligned}
f_{\mathbf{k}\mathbf{k}_1\mathbf{k}_2\mathbf{k}_3} \quad = \quad & N(\mathbf{k}_1)N(\mathbf{k}_2)N(\mathbf{k}_3) + N(\mathbf{k})N(\mathbf{k}_2)N(\mathbf{k}_3) \\
& - N(\mathbf{k})N(\mathbf{k}_1)N(\mathbf{k}_2) - N(\mathbf{k})N(\mathbf{k}_1)N(\mathbf{k}_3).
\end{aligned} \qquad (1.21)$$

Hence, the collision integral is:

$$\mathcal{I}^{(4)}\left[N(\mathbf{k})\right] = \pi \int d\mathbf{k}_1 d\mathbf{k}_2 d\mathbf{k}_3\, W_{\mathbf{k}\mathbf{k}_1\mathbf{k}_2\mathbf{k}_3}\, f_{\mathbf{k}\mathbf{k}_1\mathbf{k}_2\mathbf{k}_3}, \tag{1.22}$$

where

$$W_{\mathbf{k}\mathbf{k}_1\mathbf{k}_2\mathbf{k}_3} = |T_{\mathbf{k}\mathbf{k}_1\mathbf{k}_2\mathbf{k}_3}|^2\, \delta(\mathbf{k}+\mathbf{k}_1-\mathbf{k}_2-\mathbf{k}_3)\delta(\omega(\mathbf{k})+\omega(\mathbf{k}_1)-\omega(\mathbf{k}_2)-\omega(\mathbf{k}_3)) \tag{1.23}$$

1.3   Exact solutions of kinetic equations

The beauty of the weak turbulence theory is that for wide classes of dispersion laws it allows to obtain *exact* stationary solutions of the kinetic equations in the form of power-law Kolmogorov-like distributions. Let us first consider the **decay dispersion law**.

The first observation is that energy equipartition, i.e. Rayleigh - Jeans distribution

$$N^{RJ}(\mathbf{k}) \propto \omega(\mathbf{k})^{-1}, \tag{1.24}$$

is always solution. It is sufficient to rewrite $f_{\mathbf{k},\mathbf{k}_1,\mathbf{k}_2}$ in the form:

$$f_{\mathbf{k},\mathbf{k}_1,\mathbf{k}_2} = N(\mathbf{k})N(\mathbf{k}_1)N(\mathbf{k}_2)\left(N(\mathbf{k})^{-1}-N(\mathbf{k}_1)^{-1}-N(\mathbf{k}_2)^{-1}\right), \tag{1.25}$$

to see that due to the delta-function in ω in the collision integral $N^{RJ}(\mathbf{k})$ annihilates it as $f_{\mathbf{k},\mathbf{k}_1,\mathbf{k}_2}$ becomes proportional to $\omega(\mathbf{k})-\omega(\mathbf{k}_1)-\omega(\mathbf{k}_2)$.

Let us remind that the Rayleigh - Jeans distribution $N = \frac{T}{\omega}$ corresponds to thermodynamic equilibrium with temperature T. It is equipartition in energy because in zeroth order the mean energy density of each mode is $\epsilon(\mathbf{k}) = \omega(\mathbf{k})N(\mathbf{k})$. A generalization of this solution is a drift Rayleigh - Jeans distribution

$$N^{DRJ}(\mathbf{k}) \propto (\omega(\mathbf{k})-\mathbf{k}\cdot\mathbf{u})^{-1}, \tag{1.26}$$

which may be obtained from energy and momentum conservation. Here $\mathbf{u}$ is a vector proportional to the overall momentum of the system.

However, apart from equilibrium solutions, there are other, nonequilibrium ones with $f_{\mathbf{k},\mathbf{k}_1,\mathbf{k}_2} \neq 0$, at least for certain classes of dispersion laws. Let us suppose now that dispersion is isotropic and has the form

$$\omega(\mathbf{k}) \propto |\mathbf{k}|^{\beta} \tag{1.27}$$

(this is the case of e.g. capillary waves, see below). It is easy to see that $\beta > 1$ corresponds to a decay dispersion law. The existence of nonequilibrium solutions is based on *symmetries* of the collision integral. Indeed, dispersion law is invariant with respect to rotations in $\mathbf{k}$- space. If we denote by $\hat{r}$ rotation operator, it means that $\omega(\hat{r}\mathbf{k}) = \omega(\mathbf{k})$. The interaction coefficient $V_{\mathbf{k},\mathbf{k}_1\mathbf{k}_2}$ is a function of scalar products of the vectors $\mathbf{k}, \mathbf{k}_1, \mathbf{k}_2$ and, hence is rotation - invariant. The delta-functions are also invariant with respect to rotations. Hence, $W_{\hat{r}\mathbf{k},\hat{r}\mathbf{k}_1\hat{r}\mathbf{k}_2} = W_{\mathbf{k},\mathbf{k}_1\mathbf{k}_2}$.

Another symmetry of the dispersion law (1.27) is scale invariance. We have $\omega(\lambda \mathbf{k}) = \lambda^{\beta}\omega(\mathbf{k})$. Hence, delta- function of frequencies scales with a factor $\lambda^{-\beta}$, while delta-function of wave-numbers scales, as usual, as λ^{-d}, where d is space (and $\mathbf{k}$-space) dimension. Normally, the squares of interaction coefficients $|V|^2$ entering the collision integral are also homogeneous functions with some index m. Hence

$$W_{\lambda \mathbf{k}, \lambda \mathbf{k}_1 \lambda \mathbf{k}_2} = \lambda^{m-\beta-d} W_{\mathbf{k}, \mathbf{k}_1 \mathbf{k}_2}. \tag{1.28}$$

The triads of wave-vectors entering each of three terms in the collision integral (1.15) form three triangles with one of the sides fixed by external argument $\mathbf{k}$ ($\mathbf{k}_1, \mathbf{k}_2$ are dumb integration variables). A *crucial observation* is that by appropriate rotations and rescalings the second and the third triangles may be transformed into the first one. However, while integration measure is invariant under rotations, rescalings provide an extra (Jacobian) factor λ^{3d} which will add up to the rescalings of W (cf. (1.28)).

In this way the collision integral may be brought to the following form:

$$\mathcal{I}^{(3)}[N(\mathbf{k})] = \int d\mathbf{k}_1 d\mathbf{k}_2 \, \delta(\mathbf{k} - \mathbf{k}_1 - \mathbf{k}_2)\delta(\omega(\mathbf{k}) - \omega(\mathbf{k}_1) - \omega(\mathbf{k}_2)) \, W_{\mathbf{k}, \mathbf{k}_1 \mathbf{k}_2} \cdot$$
$$\left(f_{\mathbf{k}, \mathbf{k}_1, \mathbf{k}_2} - \lambda_1^{\alpha} f_{\hat{G}_1^2 \mathbf{k}_1, \mathbf{k}, \hat{G}_1 \mathbf{k}_2} - \lambda_2^{\alpha} f_{\hat{G}_2^2 \mathbf{k}_2, \hat{G}_2 \mathbf{k}_1, \mathbf{k}} \right). \tag{1.29}$$

Here the scaling factors are $\lambda_{1,2} = \frac{|\mathbf{k}|}{|\mathbf{k}_{1,2}|}$, and the transformations $\hat{G}_{1,2}$ are defined by rotations and rescalings with $\lambda_{1,2}$ such that $\hat{G}_{1,2}\mathbf{k}_{1,2} = \mathbf{k}$. The scaling factor is $\alpha = m + 2d - \beta$.

If one looks for power-law solutions $N(\mathbf{k}) \propto \omega^s(\mathbf{k})$, then they transform under rescaling as $N(\lambda \mathbf{k}) = \lambda^{\beta s}$, and hence $f_{\lambda \mathbf{k}, \lambda \mathbf{k}_1, \lambda \mathbf{k}_2} = \lambda^{2\beta s} f_{\mathbf{k}, \mathbf{k}_1, \mathbf{k}_2}$. In addition, f is invariant with respect to rotations in this case. Inverting the dispersion relation $|\mathbf{k}| \propto \omega^{\frac{1}{\beta}}$ and introducing this expression into the definitions of $\lambda_{1,2}$ one reduces the collision integral to the following expresion:

$$\mathcal{I}^{(3)}[N(\mathbf{k})] = \omega^{\nu}(\mathbf{k}) \int d\mathbf{k}_1 d\mathbf{k}_2 \, W_{\mathbf{k}, \mathbf{k}_1 \mathbf{k}_2} \, f_{\mathbf{k}, \mathbf{k}_1, \mathbf{k}_2} \left(\omega^{-\nu}(\mathbf{k}) - \omega^{-\nu}(\mathbf{k}_1) - \omega^{-\nu}(\mathbf{k}_2) \right), \tag{1.30}$$

where

$$\nu = 2s - 1 + \frac{2d + m}{\beta}. \tag{1.31}$$

Note that

$$f_{\mathbf{k}, \mathbf{k}_1, \mathbf{k}_2} = (\omega(\mathbf{k})\omega(\mathbf{k}_1)\omega(\mathbf{k}_2))^s \left(\omega^{-s}(\mathbf{k}) - \omega^{-s}(\mathbf{k}_1) - \omega^{-s}(\mathbf{k}_2) \right). \tag{1.32}$$

There are two solutions for s which annihilate the collision integral and, hence, give stationary distributions. At $s = -1$ f vanishes: this is the equilibrium solution. But there is another one: $\nu = -1$ and $s = -\frac{2d+m}{2\beta}$ giving

$$N(\mathbf{k}) \propto |\mathbf{k}|^{-\frac{2d+m}{2}}. \tag{1.33}$$

In distinction with the equilibrium spectrum it is non-universal and is determined by the scaling exponent of the interaction coefficients in the Hamiltonian.

The same approach may be applied to the **non-decay case**. The resonant wave *quadrangle* $(\mathbf{k}, \mathbf{k}_1, \mathbf{k}_2, \mathbf{k}_3)$ defining the integration measure in the collision integral (1.22) may be transformed, with the help of symmetry transformations $\hat{G}_i : \hat{G}_i \mathbf{k}_i = \mathbf{k}\ i = 1, 2, 3$ including rotations and dilatations, into another resonant quadrangle $(\mathbf{k}, \mathbf{q}_1, \mathbf{q}_2, \mathbf{q}_3)$ in three different ways:

$$q_1 \;=\; \hat{G}_1^{\,2}\mathbf{k}_1, \; \mathbf{q}_2 = \hat{G}_1 \mathbf{k}_2, \; \mathbf{q}_3 = \hat{G}_1 \mathbf{k}_3; \tag{1.34}$$

$$q_1 \;=\; \hat{G}_2 \mathbf{k}_3, \; \mathbf{q}_2 = \hat{G}_2^{\,2}\mathbf{k}_2, \; \mathbf{q}_3 = \hat{G}_2 \mathbf{k}_1; \tag{1.35}$$

$$q_1 \;=\; \hat{G}_3 \mathbf{k}_2, \; \mathbf{q}_2 = \hat{G}_3 \mathbf{k}_1, \; \mathbf{q}_3 = \hat{G}_3^{\,2}\mathbf{k}_3. \tag{1.36}$$

For power-law dispersion $\omega(\mathbf{k}) \propto |\mathbf{k}|^\beta$ the interaction coefficients and delta-function are scale-invariant, similar to the decay-spectrum case and the collision integral may be represented as a sum of four replicas of itself in the form:

$$\mathcal{I}^{(4)}\left[N(\mathbf{k})\right] = \frac{1}{4}\int d\mathbf{k}_1 d\mathbf{k}_2 d\mathbf{k}_3\, W_{\mathbf{k}\mathbf{k}_1\mathbf{k}_2\mathbf{k}_3}\left(f_{\mathbf{k}} + \lambda_1^\alpha f_{\hat{G}_1\mathbf{k}} + \lambda_2^\alpha f_{\hat{G}_2\mathbf{k}} + \lambda_3^\alpha f_{\hat{G}_3\mathbf{k}}\right), \tag{1.37}$$

where the scale factors are $\lambda_i = \frac{|\mathbf{k}|}{|\mathbf{k}_i|}$, $i = 1, 2, 3$, and $\alpha = m + 3d - \beta$. As a result, the collision integral, for isotropic and homogeneous distributions $N(\mathbf{k}) \propto \omega^s(\mathbf{k})$, is factorized:

$$\mathcal{I}^{(4)}\left[N(\mathbf{k})\right] \;=\; \frac{\omega^\nu(\mathbf{k})}{4}\int d\mathbf{k}_1 d\mathbf{k}_2\, d\mathbf{k}_3\, W_{\mathbf{k},\mathbf{k}_1\mathbf{k}_2,\mathbf{k}_3}\, f_{\mathbf{k},\mathbf{k}_1,\mathbf{k}_2,\mathbf{k}_3}\,\cdot$$
$$\left(\omega^{-\nu}(\mathbf{k}) + \omega^{-\nu}(\mathbf{k}_1) - \omega^{-\nu}(\mathbf{k}_2)\omega^{-\nu}(\mathbf{k}_3)\right), \tag{1.38}$$

where

$$\nu = 3s - 1 + \frac{3d + m}{\beta} \tag{1.39}$$

and

$$f_{\mathbf{k},\mathbf{k}_1,\mathbf{k}_2,\mathbf{k}_3} \;=\; (\omega(\mathbf{k})\omega(\mathbf{k}_1)\omega(\mathbf{k}_2)\omega(\mathbf{k}_2))^s \left(\omega^{-s}(\mathbf{k}) + \omega^{-s}(\mathbf{k}_1)\right)$$
$$-\;\omega^{-s}(\mathbf{k}_2) - \omega^{-s}(\mathbf{k}_2)). \tag{1.40}$$

Hence, the collision integral vanishes, i.e. stationary distributions result, either when f vanishes, or when the last factor vanishes. The first case corresponds to two limiting cases $\mu \gg \omega$ and $\mu \ll \omega$ of the equilibrium distribution with "chemical potential" μ:

$$N^{RJ}(\mathbf{k}) \propto (\omega(\mathbf{k}) - \mu)^{-1}, \tag{1.41}$$

which, as in the decay case, may be extended to include a drift:

$$N^{DRJ}(\mathbf{k}) \propto (\omega(\mathbf{k}) - \mu - \mathbf{k} \cdot \mathbf{u})^{-1}. \tag{1.42}$$

The appearance of chemical potential is related to additional conservation law of the number of "particles" in non-decay kinetic equations, see below.

The non-equilibrium stationary distributions correspond to $\nu = 0, -1$ and are related to conservation of number of waves and energy, respectively, in an elementary collision proccess. Correspondingly, $s = \frac{1}{3} - \frac{3d+m}{3\beta}$, and $s = -\frac{3d+m}{3\beta}$ which gives two solutions for N:

$$N(\mathbf{k}) \propto |\mathbf{k}|^{\frac{\beta - 3d + m}{3}}, \tag{1.43}$$

$$N(\mathbf{k}) \propto |\mathbf{k}|^{-\frac{3d+m}{3}} \tag{1.44}$$

1.4   Conservation laws and dimensional estimates

The *mean energy* of the medium in the lowest order (cf. (1.5)) is $E = \int d\mathbf{k}\,\omega(\mathbf{k})N(\mathbf{k})$, whence $\epsilon = \omega(\mathbf{k})N(\mathbf{k})$ is the mean energy density. Multiplying both sides of (1.10) by $\omega(\mathbf{k})$, integrating by $\mathbf{k}$, and using the symmetry properties of interaction coefficients we get, e.g. in the decay spectrum case:

$$\dot{E} = \int d\mathbf{k}d\mathbf{k}_1 d\mathbf{k}_2\,(\omega(\mathbf{k}) - \omega(\mathbf{k}_1) - \omega(\mathbf{k}_2)) \cdot$$
$$2\pi\,\,|V_{\mathbf{k}\mathbf{k}_1\mathbf{k}_2}|^2\,\delta(\mathbf{k} - \mathbf{k}_1 - \mathbf{k}_2)\delta(\omega(\mathbf{k}) - \omega(\mathbf{k}_1) - \omega(\mathbf{k}_2)) \equiv 0, \tag{1.45}$$

and similarly in the decay case. Thus, the total energy is conserved by the kinetic equation. It is clear from the above expression that total energy conservation is ensured by energy conservation in each elementary interaction act.

The conservation law may rewritten in (pseudo-) local form:

$$\frac{\partial \epsilon}{\partial t} + \frac{\partial \mathbf{P}}{\partial \mathbf{k}} = 0, \tag{1.46}$$

where $\mathbf{P}$ is (the density of) the energy flux in the $\mathbf{k}$- space: $\frac{\partial \mathbf{P}}{\partial \mathbf{k}} = -\omega(\mathbf{k})\mathcal{I}\,[N(\mathbf{k})]$. The true locality is ensured if integrals defining P are convergent (*locality criterion*).

It may be shown in analogous way that the three components of the *total momentum* of the medium $\mathbf{K} = \int d\mathbf{k}\,\mathbf{k}N(\mathbf{k})$ are conserved by the kinetic equation, as they are conserved in each elementary interaction.

Moreover, in the non-decay case the total *wave action*, or the total number of waves (quasi-particles) $N = \int d\mathbf{k}N(\mathbf{k})$ is also conserved, as it is conserved in elementary interactions. It may be again rewritten in the (pseudo-) local form:

$$\frac{\partial N(\mathbf{k})}{\partial t} + \frac{\partial \mathbf{Q}}{\partial \mathbf{k}} = 0, \tag{1.47}$$

where $\mathbf{Q}$ is (the density of) the wave-action flux in the $\mathbf{k}$- space: $\frac{\partial \mathbf{Q}}{\partial \mathbf{k}} = -\mathcal{I}^{(4)}\,[N(\mathbf{k})]$, the locality being ensured if integrals converge.

Like the famous Kolmogorov - Obukhov spectrum of (strong) hydrodynamical turbulence, energy spectra corresponding to constant fluxes of conserved quantities may be constructed by dimensional considerations. The difference between isotropic and homogeneous developed turbulence and wave turbulence is that in the last case additional parameter - frequency of the waves which constitute the turbulent field - is present. Hence, pure dimensional considerations are insufficient and some dynamical information should be added. It is provided by the very structure of the kinetic equation with either $\mathcal{I}^{(3)} \propto N^2$, or $\mathcal{I}^{(4)} \propto N^3$. In its turn the collision integral defines the conserved quantity flux, as just explained. The dimensions of N, P, and Q are:

$$[N] = L^5 T^{-1}, \quad [P] = L^{5-d} T^{-3}, \quad [Q] = L^{5-d} T^{-2}. \tag{1.48}$$

In the decay-spectrum case $N \sim P^{\frac{1}{2}}$ and looking for the spectrum of the form $N = P^{\frac{1}{2}} \omega^a |\mathbf{k}|^b$ we get by comparing dimensions:

$$N \propto P^{\frac{1}{2}} \omega^{-\frac{1}{2}} |\mathbf{k}|^{-\frac{d+5}{2}}. \tag{1.49}$$

Under hypothesis of full self-similarity (necessary for applying dimensional arguments), the interaction coefficients should scale as some combination of wavenumber and frequency (no P). Hence $|V_{\mathbf{k}\mathbf{k}_1 \, bk_2}|^2 \sim |\mathbf{k}|^{5-d} \omega \Rightarrow m = 5 - d + \beta$ and we thus recover the earlier result (1.33).

In the non-decay spectrum case $N \sim P^{\frac{1}{3}}$ or $N \sim Q^{\frac{1}{3}}$ and we get a spectrum with a constant energy flux:

$$N \propto P^{\frac{1}{3}} |\mathbf{k}|^{-\frac{d+10}{3}}, \tag{1.50}$$

and that with a constant wave-action flux:

$$N \propto Q^{\frac{1}{3}} \omega^{\frac{1}{3}} |\mathbf{k}|^{-\frac{d+10}{3}}. \tag{1.51}$$

In case of full self-similarity these spectra correspond to those found earlier.

1.5 Application to surface waves

Waves on the free surface of the fluid subject to capillary forces are described in terms of the position of fluid surface $\eta(\mathbf{x}, t)$ and the value of the fluid velocity potential $\phi(\mathbf{x}, t)$ (assumption of potential flow is made) at the free surface: $\psi = \phi|_{z=\eta}$. The Hamiltonian of the system is given by energy per unit mass:

$$H = \frac{1}{2} \int d\mathbf{x} \int_{-\infty}^{\eta} dz \, (\nabla \phi)^2 + \frac{g}{2} \int d\mathbf{x} \, \eta^2 + \frac{\alpha}{\rho} \int d\mathbf{x} \left(\sqrt{1 + (\nabla \eta)^2} - 1 \right). \tag{1.52}$$

Here g is gravity acceleration, α is the surface tension coefficient, ρ is the fluid density. Normal variables a, a^* are introduced via the Fourier transforms of η and ψ:

$$\eta(\mathbf{k}) = \frac{1}{2\pi} \sqrt{\frac{|\mathbf{k}|}{2\omega(\mathbf{k})}} \left(a_{\mathbf{k}} + a^*_{-\mathbf{k}} \right), \quad \psi(\mathbf{k}) = -\frac{i}{2\pi} \sqrt{\frac{\omega(\mathbf{k})}{2|\mathbf{k}|}} \left(a_{\mathbf{k}} - a^*_{-\mathbf{k}} \right). \tag{1.53}$$

The dispersion relation for capillary-gravity waves is:

$$\omega(\mathbf{k}) = \sqrt{g|\mathbf{k}| + \frac{\alpha}{\rho} |\mathbf{k}|^3}. \tag{1.54}$$

The second derivative of the dispersion curve changes sign at $k_0 = \sqrt{\frac{\rho g}{\alpha}}$. The whole dispersion law is not self-similar, so we will consider separately capillary waves, $k \ll k_0$, and gravity waves (deep fluid), $k \gg k_0$.

For **capillary waves** $\omega(\mathbf{k}) \propto |\mathbf{k}|^{\frac{3}{2}}$, hence $\beta = \frac{3}{2}$ and $d = 2$. A stationary spectrum, thus results:

$$N \propto P^{\frac{1}{2}} |\mathbf{k}|^{-\frac{17}{4}}. \tag{1.55}$$

For **gravity waves** $\omega(\mathbf{k}) \propto |\mathbf{k}|^{\frac{1}{2}}$, hence $\beta = \frac{1}{2}$ and $d = 2$. A stationary spectrum with constant P is

$$N \propto P^{\frac{1}{3}} |\mathbf{k}|^{-4}, \tag{1.56}$$

and that with constant Q is

$$N \propto Q^{\frac{1}{3}} |\mathbf{k}|^{-\frac{23}{6}}. \tag{1.57}$$

1.6 Historical comments and bibliography

The idea of weak turbulence appeared first in the context of plasma physics. The whole group led by Sagdeev who was, probably, the first to promote it, worked on different applications of this idea in early and mid-sixties in Novosibirsk: Sagdeev and Galeev (1969) , Galeev and Karpman (1963) , Zaslavsky and Sagdeev (1967) . Zakharov (cf. Zakharov (1984) and references therein) was first to apply these ideas to hydrodynamical waves. Above-displayed power spectra of capillary-gravity waves were obtained in the pioneering paper Zakharov and Filonenko (1966) . At the same time Hasselmann developed statistical approach to water waves Hasselman (1967) .

Kinetic equation was derived in the above-cited works, where approximations leading to it were also extensively discussed. At the same time random-wave closures in the hydrodynamical context were otained by Benney and Saffman (1966) and Benney and Newell (1969) .

Above, we followed heuristic quantum-mechanical short-cut in order to derive the kinetic equation. For direct derivation see, e.g. Zakharov, Lvov and Falkovich (1992) . For a systematic derivation via phase averaging and coarse-graining procedure in action-angle variables see the work of Zaslavsky and Sagdeev (1967) .

The presentation above is standard. There is a number of books and reviews on the subject: Zakharov (1984) , Zakharov, Lvov and Falkovich (1992) , Kadomtsev and Kontorovich (1974) . Zakharov (1984) used transformations of frequency variables in the collision integral in order to obtain exact stationary solutions. We followed above an alternative approach developed by Kats and Kontorovich (1973) , which is also applicable for non strictly self-similar distributions (see below). For discussion of applicability criteria of kinetic equations, see Zakharov, Lvov and Falkovich (1992) as well as for extensive discussion of stability and locality of exact power-law solutions. The list of references presented of the end of the chapter is, obviously, non-exhaustive.

2 Weak turbulence of waves in the rotating shallow water model

2.1 Rotating shallow water model (RSW): a reminder

Physical meaning of the model: horizontal momentum and mass conservation for a layer of shallow water in hydrostatic approximation on the tangent plane to a rotating planet. Rotation: Coriolis force. Centrifugal force is included in the effective gravity g. Stratification is rudimentary and reduced to the dynamics of a single isopycnal surface

$z = h(x, y, t)$, the free surface. Fluid at rest: $h = h_0$. Equations of motion (non-dissipative):

$$\partial_t \mathbf{v} + \mathbf{v} \cdot \nabla \mathbf{v} + f(y)\hat{\mathbf{z}} \wedge \mathbf{v} + g\nabla h = 0, \qquad (2.1)$$

$$\partial_t h + \nabla \cdot (\mathbf{v}h) = 0, \qquad (2.2)$$

Coriolis parameter: $f(y)$ (the meridional dependence of the normal component of the planet's angular velocity is the only remnant of the planet's sphericity).
  • Mid-latiude tangent plane: $f = f_0 + \beta y$.
  • Equatorial tangent plane: $f = \beta y$.
If sphericity is neglected, $f = f_0 = const$: f - plane approximation.
  Key quantity: potential vorticity (PV):

$$q = \frac{v_x - u_y + f(y)}{h}, \qquad (2.3)$$

Here $v_x - u_y$ is relative vorticity, $v_x - u_y + f(y)$ is total (relative plus planetary $f(y)$) vorticity. PV is a Lagrangian invariant:

$$\frac{dq}{dt} = 0, \quad \frac{d}{dt} = \partial_t + \mathbf{v} \cdot \nabla. \qquad (2.4)$$

Waves in the RSW are obtained by linearization. Linearization of (2.1) about the rest state on the f - plane results in a system of linear PDE with constant coefficients. Looking for solutions $\propto e^{i(\mathbf{k} \cdot \mathbf{x} - \omega t)}$ results in the dispersion relation:

$$\omega \left(\omega^2 - gh_0 \mathbf{k}^2 - f^2 \right) = 0. \qquad (2.5)$$

The root $\omega = 0$ corresponds to linearized PV, which is a *slow variable*. Two other roots correspond to the *fast variables*, surface inertia-gravity waves (SIGW). Specifics of SIGW: 1) spatial isotropy, 2) spectral gap (no waves with $\omega < f_0$), 3) absence of dispersion for short ($k \gg 1$) waves.

  The presence of both slow and fast variables in the system is typical for ocean and atmosphere motions.

  For small *Rossby numbers* $Ro = \frac{U}{fL}$, where U and L are typical velocity and horizontal scale of the motion, slow and fast variables are *dynamically decoupled* and, for not too long times, may be treated independently (for proof see Reznik, Zeitlin and Ben Jelloul (2001)). A closed equation arises for slow evolution of PV which, for small Ro, may be expressed in terms of h via *geostrophic balance* between the pressure force and the Coriolis force:

$$\partial_t \left(h - \nabla^2 h \right) - \mathcal{J} \left(h, \nabla^2 h \right) = 0. \qquad (2.6)$$

Here the Rossby deformation radius $R_d = \frac{\sqrt{gh_0}}{f^2}$, an intrinsic length-scale present in the system, is taken as a length unit, and the time - scale is L/U in contradistinction with the fast time-scale f^{-1} of SIGW.

  There are no waves in (2.6). They, however, appear if the slow dynamics is considered on the β - plane. In this case, supposing non-dimensional β to be of the same order of magnitude as Ro, one obtains instead of (2.6) the following equation:

$$\partial_t \left(h - \nabla^2 h \right) - \beta \partial_x h - \mathcal{J} \left(h, \nabla^2 h \right) = 0. \qquad (2.7)$$

(we keep β to indicate the origin of the last term). This equation, if linearized, produces *Rossby waves* with dispersion relation:

$$\omega = -\beta \frac{k_x}{\mathbf{k}^2 + 1}, \tag{2.8}$$

where k_x is the wave vector component in x- direction. The most characteristic features of Rossby waves are: unique sign of phase velocity in the x- direction, strong spatial anisotropy and non-monotonicity of the dispersion curve.

Finally, on the *equatorial β - plane* with boundary conditions of exponential decay far from the equator (equatorial wave-guide) specific hybrid wave motions appear. They are obtained from linearization of (2.1) and decomposition of all fields in parabolic cylinder functions of the form

$$\phi_n(y) = \frac{H_n(y)e^{-\frac{y^2}{2}}}{\sqrt{2^n n! \sqrt{\pi}}}, \tag{2.9}$$

where H_n are Hermite polynomials. They are (in non-dimensional units):
- Kelvin waves with linear dispersion $\omega = k$,
- Yanai waves with the dispersion law $\omega^2 - k\omega - 1 = 0$,
- Rossby and inerta-gravity waves with the dispersion law $\omega^3 - (k^2 + (2n+1))\omega - k = 0$ (lower frequency - Rossby wave, higher frequencies - inertia-gravity waves).

Thus, fast SIGW, both at mid-latitudes and at equator, are strongly dispersive at long wave-lengths and weakly dispersive at short wave-lengths. Rossby waves are strongly-dispersive and strongly spatially anisotropic. Equatorial Kelvin waves are rigorously non-dispersive. Short Yanai waves join Rossby-waves family for negative k_x and SIGW family for positive k_x (see figure). Below we will show how the weak turbulence approach may be applied to waves in RSW. We will treat each type of waves separately (SIGW on the f- plane, equatorial SIGW, Rossby waves on the β- plane).

2.2 Factorization of the collision integral for almost non-dispersive waves

The dispersion relation (2.5) is, obviously, not scale-invariant, so the standard recipies of the weak turbulence do not apply. However, for dispersion laws of the form:

$$\omega(\mathbf{k}) = c|\mathbf{k}| + \gamma(\mathbf{k}), \tag{2.10}$$

considered in the domains of $\mathbf{k}$ such that $|\gamma| \ll c|\mathbf{k}|$, the method of conformal transformations of the integration domain in the collision integral may be extended to get, under some additional hypotheses, factorization. Indeed, the main spectrum being of non-decay type, the collision integral is of the form (1.22). The smallness of dispersion means that wave scattering is dominated by small-angle processes. For angles between the wave-vectors such that

$$\theta_i = \widehat{(\mathbf{k}, \mathbf{k}_i)} \le \sqrt{\frac{\gamma}{c|\mathbf{k}|}} \tag{2.11}$$

the following estimate then holds for the four-wave interaction coefficient, cf (1.19): $T \sim \frac{V^2}{c|\mathbf{k}|(\theta^2 + |\gamma/c|\mathbf{k}||)}$ and everywhere, except for resonance denominators and energy conservation, one may take $\mathbf{k}_i$ parallel to $\mathbf{k}$ (if V does not vanish at such values of $\mathbf{k}_i$).

Introducing the variables $s_i = \frac{k_{i\perp}}{|k_{i\perp}|}$ the product of δ- functions in the collision integral (1.22), (1.23) may be then rewritten as

$$\delta(|\mathbf{k}| + |\mathbf{k}_1| - |\mathbf{k}_2| - |\mathbf{k}_3|)\, \delta(s_1|\mathbf{k}|_1\theta_1 - s_2|\mathbf{k}|_2\theta_2 - s_3|\mathbf{k}|_3\theta_3)$$
$$\delta\left(\gamma + \gamma_1 - \gamma_2 - \gamma_3 + \frac{c}{2}\left(|\mathbf{k}|_1\theta_1^2 - |\mathbf{k}|_2\theta_2^2 - |\mathbf{k}|_3\theta_3^2\right)\right). \tag{2.12}$$

Collision integral is not conformal-invariant. Nevertheless, if

$$\gamma(\lambda\mathbf{k}) = \lambda^\beta \gamma(\mathbf{k}), \; V_{\lambda\mathbf{k},\lambda\mathbf{k}_1,\lambda\mathbf{k}_2} = \lambda^\mu V_{\mathbf{k},\mathbf{k}_1,\mathbf{k}_2}, \tag{2.13}$$

then the collision integral may be factorized using rescaling of angles and conformal transformations

Indeed, in this case the conservation laws and resonant denominators are mapped onto themselves under the transformation $\mathbf{k}_i \to \lambda\mathbf{k}_i, \; \theta_i \to \lambda^{\frac{\beta-1}{2}}\theta_i$. Correspondingly:

$$W_{\mathbf{k},\mathbf{k}_1,\mathbf{k}_2,\mathbf{k}_3} \to \lambda^w W_{\mathbf{k},\mathbf{k}_1,\mathbf{k}_2,\mathbf{k}_3}, \; w = 4\mu - 3\beta - 1 - \frac{1}{2}(\beta+1)(d-1). \tag{2.14}$$

If one supposes that distribution functions are homogeneous in $|\mathbf{k}|$ (with possible slow angle dependence): $N(\mathbf{k}) \propto |\mathbf{k}|^s$, then the collision integral may be factorized by the following transformations (the scaling factors λ_i are defined as before and non-bold notation is used for wave-vector moduli):

$$k_2' = \lambda_2 k = \lambda_2^2 k_2, \; k_1' = \lambda_2 k_3, \; k_3' = \lambda_2 k_1, \tag{2.15}$$

$$\theta_2' = -\lambda_2^{\frac{\beta-1}{2}}\theta_2, \; \theta_1' = \lambda_2^{\frac{\beta-1}{2}}(\theta_3 - \theta_2), \; \theta_3' = \lambda_2^{\frac{\beta-1}{2}}(\theta_1 - \theta_2), \tag{2.16}$$

and the collision integral becomes:

$$\mathcal{I}^{(4)}[N(\mathbf{k})] = \frac{k^\nu}{4}\int d\mathbf{k}_1 d\mathbf{k}_2 \, d\mathbf{k}_3 \, W_{\mathbf{k},\mathbf{k}_1\mathbf{k}_2,\mathbf{k}_3} \, f_{\mathbf{k},\mathbf{k}_1,\mathbf{k}_2,\mathbf{k}_3} \cdot$$
$$\left(k^{-\nu} + k_1^{-\nu} - k_2^{-\nu} - k_3^{-\nu}\right), \tag{2.17}$$

where

$$\nu = 3s + w + 4d + \frac{3}{2}(d-1)(\beta-1). \tag{2.18}$$

Hence, there are two non-equilibrium power-law spectra corresponding to $\nu = 0$ (constant wave- action flux solution), and to $\nu = -1$ (constant energy flux solution).

2.3 Weak turbulence of short SIGW on the f- plane

As was already mentioned, RSW equations describe both slow vortex (zero frequency in the linear approximation) and fast SIGW motions. An invariant criterion of their separation, at least at weak nonlinearities, is provided by the PV. Indeed, SIGW do not bear PV - anomaly, i.e. deviation of PV from the background value $\frac{f}{h_0}$. For zero PV-anomaly, i.e. for *constant PV*, the motion is described by a pair of canonical Hamiltonian variables: velocity potential Φ and h. Namely for two components of velocity in this case we have:

$$u = \frac{\partial \Phi}{\partial x} - \frac{f}{h_0}\frac{\partial}{\partial y}\Delta^{-1}(h - h_0), \quad v = \frac{\partial \Phi}{\partial y} + \frac{f}{h_0}\frac{\partial}{\partial x}\Delta^{-1}(h - h_0), \tag{2.19}$$

where Δ^{-1} denotes inverse Laplacian. Thus reduced system is Hamiltonian

$$\dot{h} = \frac{\delta H}{\delta \Phi}, \quad \dot{\Phi} = -\frac{\delta H}{\delta h}, \tag{2.20}$$

with a Hamiltonian given by the full (kinetic + potential) energy:

$$H = \frac{1}{2}\int dx\, dy\, \left[h(u^2 + v^2) + gh^2\right]. \tag{2.21}$$

Introducing normal wave amplitudes $b_\mathbf{k}$ such that

$$h_\mathbf{k} = \sqrt{\frac{k^2 h_0}{2\omega(\mathbf{k})}}(b_{-\mathbf{k}}^* - b_{-\mathbf{k}}), \quad \Phi_\mathbf{k} = \sqrt{\frac{\omega(\mathbf{k})}{2k^2 h_0}}(b_{-\mathbf{k}}^* - b_{-\mathbf{k}}), \tag{2.22}$$

we get the interaction Hamiltonian of the form:

$$\begin{aligned} H_3 &= \frac{1}{2}\int dk dk_1 dk_2\, V_{\mathbf{k}\mathbf{k}_1\mathbf{k}_2} b_\mathbf{k}^* b_{\mathbf{k}_1} b_{\mathbf{k}_2} + c.c. \\ &+ \frac{1}{3}\int dk dk_1 dk_2\, U_{\mathbf{k}\mathbf{k}_1\mathbf{k}_2} b_\mathbf{k} b_{\mathbf{k}_1} b_{\mathbf{k}_2} + c.c. \end{aligned} \tag{2.23}$$

The interaction coefficients are:

$$V_{\mathbf{k}\mathbf{k}_1\mathbf{k}_2} = 2U_{\mathbf{k}\mathbf{k}_1\mathbf{k}_2} = \sqrt{18g}\frac{\mathbf{k}_2\cdot\mathbf{k}_3\,\omega(\mathbf{k}_1) + \mathbf{k}_1\cdot\mathbf{k}_3\,\omega(\mathbf{k}_2) + \mathbf{k}_1\cdot\mathbf{k}_2\,\omega(\mathbf{k}_3)}{\sqrt{\omega(\mathbf{k}_1)\omega(\mathbf{k}_2)\omega(\mathbf{k}_3)}}. \tag{2.24}$$

The presence of "vacuum-to-three-particles" decay terms with coefficients U in (2.23) is not dangerous as *there are no three-wave resonances in RSW on the f- plane.* By this reason further canonical transformation

$$\begin{aligned} b_\mathbf{k} \rightarrow\ & b_\mathbf{k} - \int dk_1 dk_2\, \delta(\mathbf{k} - \mathbf{k}_1 - \mathbf{k}_2)\frac{V_{\mathbf{k}\mathbf{k}_1\mathbf{k}_2}}{\omega(\mathbf{k}) - \omega(\mathbf{k}_1) - \omega(\mathbf{k}_2)} b_{\mathbf{k}_1} b_{\mathbf{k}_2} \\ & -2\int dk_1 dk_2\, \delta(\mathbf{k} + \mathbf{k}_1 - \mathbf{k}_2)\frac{V_{\mathbf{k}\mathbf{k}_1\mathbf{k}_2}}{-\omega(\mathbf{k}) + \omega(\mathbf{k}_1) - \omega(\mathbf{k}_2)} b_{\mathbf{k}_1}^* b_{\mathbf{k}_2} \\ & -\int dk_1 dk_2\, \delta(\mathbf{k} + \mathbf{k}_1 + \mathbf{k}_2)\frac{U_{\mathbf{k}\mathbf{k}_1\mathbf{k}_2}}{\omega(\mathbf{k}) + \omega(\mathbf{k}_1) + \omega(\mathbf{k}_2)} b_{\mathbf{k}_1} b_{\mathbf{k}_2} \end{aligned} \tag{2.25}$$

is necessary to bring the Hamiltonian to the normal form:

$$\begin{aligned} H &= \int dk\omega(\mathbf{k})\, a_\mathbf{k} a_\mathbf{k}^* \\ &+ \frac{1}{2}\int dk dk_1 dk_2 dk_3\, \delta(\mathbf{k} + \mathbf{k}_1 - \mathbf{k}_2 - \mathbf{k}_3)\, T_{\mathbf{k}\mathbf{k}_1\mathbf{k}_2\mathbf{k}_3}\, a_\mathbf{k}^* a_{\mathbf{k}_1}^* a_{\mathbf{k}_2} a_{\mathbf{k}_3}. \end{aligned} \tag{2.26}$$

There was no generality loss up to now. However, the dispersion law (2.5) is not scale-invariant and finding exact power-law solutions of the four-wave kinetic equation is therefore impossible. We consider, hence, the short SIGW with $|\mathbf{k}| \to \infty$ and retain only the leading term in the expansion of the dispersion law in powers of $|\mathbf{k}|^{-1}$:

$$\omega(\mathbf{k}) = \sqrt{gh_0}|\mathbf{k}| \left(1 + \frac{1}{2\left(|\mathbf{k}|R_d\right)^2} \right). \tag{2.27}$$

This dispersion law is exactly of the form (2.10) with scale-invariant addition to the "acoustic" dispersion law $\omega = ck$. Hence, the theory of the previous subsection may be applied, which gives two power-law spectra:

$$N(\mathbf{k}) \propto |\mathbf{k}|^{-\frac{14}{3}}, \quad N(\mathbf{k}) \propto |\mathbf{k}|^{-\frac{13}{3}}, \tag{2.28}$$

corresponding to constant energy and wave-action fluxes through the spectrum, respectively.

2.4   Weak turbulence of short SIGW on the equatorial β- plane

The non-dimensional equations for velocities u, v and deviation of the height field from the rest value $z = \frac{h-h_0}{h_0}$ on the equatorial beta - plane are:

$$u_t + uu_x + vu_y - \beta yv + z_x = 0, \tag{2.29}$$

$$v_t + uv_x + vv_y + \beta yu + z_y = 0, \tag{2.30}$$

$$z_t + u_x + v_y + (uz)_x + (vz)_y = o. \tag{2.31}$$

For simplicity we suppose that there is unique small parameter ϵ, and assume a weak nonlinearity $u, v, z \sim \epsilon$ and a weak inhomogeneity $\beta \sim \epsilon$. Smallness of the β - term means, in fact, that we are considering motions with a characteristic scale small with respect to the equatorial deformation radius $R_e = \frac{(gh_0)^{\frac{1}{4}}}{\sqrt{\beta}}$.

Thus the leading- order part of the system (2.29) - (2.31) is a system with constant coefficients

$$u_t + z_x = 0, \tag{2.32}$$

$$v_t + z_y = 0, \tag{2.33}$$

$$z_t + u_x + v_y = 0. \tag{2.34}$$

Introducing the Fourier - transforms:

$$(u(\mathbf{r}), v(\mathbf{r}), z(\mathbf{r})) = \int (u_\mathbf{k}, v_\mathbf{k}, z_\mathbf{k}) \exp(i k r) dk, \tag{2.35}$$

where $\mathbf{r} = (x, y)$ and $\mathbf{k} = (k_1, k_2)$, and integration is over the whole plane, we rewrite the equations (2.29) - (2.31) for Fourier-transforms in a symmetric form:

$$
\begin{pmatrix} \partial u_{\mathbf{k}}/\partial t \\ \partial v_{\mathbf{k}}/\partial t \\ \partial z_{\mathbf{k}}/\partial t \end{pmatrix} + \begin{pmatrix} 0 & -i\beta\frac{\partial}{\partial k_2} & ik_1 \\ i\beta\frac{\partial}{\partial k_2} & 0 & ik_2 \\ ik_1 & ik_2 & 0 \end{pmatrix} \begin{pmatrix} u_{\mathbf{k}} \\ v_{\mathbf{k}} \\ z_{\mathbf{k}} \end{pmatrix} +
$$

$$
+ \begin{pmatrix} ik_1/2 \int (u_{\mathbf{l}}u_{\mathbf{m}} + v_{\mathbf{l}}v_{\mathbf{m}})d\lambda - \int \Omega_{\mathbf{l}}v_{\mathbf{m}}d\lambda \\ ik_2/2 \int (u_{\mathbf{l}}u_{\mathbf{m}} + v_{\mathbf{l}}v_{\mathbf{m}})d\lambda + \int \Omega_{\mathbf{l}}u_{\mathbf{m}}d\lambda \\ ik_1 \int z_{\mathbf{l}}u_{\mathbf{m}}d\lambda + ik_2 \int z_{\mathbf{l}}v_{\mathbf{m}}d\lambda \end{pmatrix} = 0,
$$

(2.36)

where $\Omega = v_x - u_y$, $\Omega_{\mathbf{l}} = il_1v_{\mathbf{l}} - il_2u_{\mathbf{l}}$, $d\lambda = \delta(\mathbf{k} - \mathbf{l} - \mathbf{m})d\mathbf{l}\,d\mathbf{m}$.

We diagonalize the main part by a change of variables

$$
\begin{pmatrix} u_{\mathbf{k}} \\ v_{\mathbf{k}} \\ z_{\mathbf{k}} \end{pmatrix} = \begin{pmatrix} \frac{-ik_2}{|\mathbf{k}|} & \frac{k_1}{\sqrt{2}|\mathbf{k}|} & \frac{-k_1}{\sqrt{2}|\mathbf{k}|} \\ \frac{ik_1}{|\mathbf{k}|} & \frac{k_2}{\sqrt{2}|\mathbf{k}|} & \frac{-k_2}{\sqrt{2}|\mathbf{k}|} \\ 0 & \frac{1}{\sqrt{2}} & \frac{1}{\sqrt{2}} \end{pmatrix} \begin{pmatrix} a_{\mathbf{k}} \\ b_{\mathbf{k}} \\ c_{\mathbf{k}} \end{pmatrix}.
$$

(2.37)

From (2.37) for real-valued u, v, z we have

$$
a_{\mathbf{k}} = \bar{a}_{-\mathbf{k}}, \quad c_{\mathbf{k}} = \bar{b}_{-\mathbf{k}}
$$

(2.38)

and as a result of the diagonalization we have

$$
\begin{pmatrix} \partial a_{\mathbf{k}}/\partial t \\ \partial b_{\mathbf{k}}/\partial t \\ \partial c_{\mathbf{k}}/\partial t \end{pmatrix} + \begin{pmatrix} 0 & 0 & 0 \\ 0 & i|\mathbf{k}| & 0 \\ 0 & 0 & -i|\mathbf{k}| \end{pmatrix} \begin{pmatrix} a_{\mathbf{k}} \\ b_{\mathbf{k}} \\ c_{\mathbf{k}} \end{pmatrix} +
$$

$$
+\beta \begin{pmatrix} -\frac{ik_1}{|\mathbf{k}|^2} & \frac{1}{\sqrt{2}}\frac{\partial}{\partial k_2} & -\frac{1}{\sqrt{2}}\frac{\partial}{\partial k_2} \\ \frac{1}{\sqrt{2}}\frac{\partial}{\partial k_2} & -\frac{ik_1}{2|\mathbf{k}|^2} & \frac{ik_1}{2|\mathbf{k}|^2} \\ -\frac{1}{\sqrt{2}}\frac{\partial}{\partial k_2} & \frac{ik_1}{2|\mathbf{k}|^2} & -\frac{ik_1}{2|\mathbf{k}|^2} \end{pmatrix} \begin{pmatrix} a_{\mathbf{k}} \\ b_{\mathbf{k}} \\ c_{\mathbf{k}} \end{pmatrix} + NL = 0 \quad .
$$

(2.39)

Here nonlinear terms NL have the form:

$$
\begin{pmatrix} \int (U^{(0)}_{klm}a_{\mathbf{l}}a_{\mathbf{m}} + U^{(1)}_{klm}a_{\mathbf{l}}b_{\mathbf{m}} + U^{(2)}_{klm}a_{\mathbf{l}}c_{\mathbf{m}})d\lambda \\ \int (V^{(0)}_{klm}a_{\mathbf{l}}b_{\mathbf{m}} + V^{(1)}_{klm}a_{\mathbf{l}}a_{\mathbf{m}} + V^{(2)}_{klm}a_{\mathbf{l}}c_{\mathbf{m}})d\lambda \\ \int (W^{(0)}_{klm}a_{\mathbf{l}}c_{\mathbf{m}} + W^{(1)}_{klm}a_{\mathbf{l}}a_{\mathbf{m}} + W^{(2)}_{klm}a_{\mathbf{l}}b_{\mathbf{m}})d\lambda \end{pmatrix} +
$$

$$
+ \begin{pmatrix} 0 \\ \int (V^{(3)}_{klm}b_{\mathbf{l}}b_{\mathbf{m}} + V^{(4)}_{klm}c_{\mathbf{l}}c_{\mathbf{m}} + V^{(5)}_{klm}b_{\mathbf{l}}c_{\mathbf{m}})d\lambda \\ \int (W^{(3)}_{klm}c_{\mathbf{l}}c_{\mathbf{m}} + W^{(4)}_{klm}b_{\mathbf{l}}b_{\mathbf{m}} + W^{(5)}_{klm}c_{\mathbf{l}}b_{\mathbf{m}})d\lambda \end{pmatrix} = 0,
$$

(2.40)

with interaction coefficients which can be easily found from (2.37).

Thus, the variable $a_{\mathbf{k}}$ describes the short equatorial Rossby waves with a dispersion law $\Omega_{\mathbf{k}} = -\frac{\beta k_1}{|\mathbf{k}|^2}$ and the variables $b_{\mathbf{k}}$, $c_{\mathbf{k}}$ describe the short inertia-gravity waves with a dispersion law $\omega_{\mathbf{k}} = |\mathbf{k}| - \frac{\beta k_1}{2|\mathbf{k}|^2}$.

The linear part of (2.39) may be diagonalized by further transforming the variables:

$$
\begin{pmatrix} a_{\mathbf{k}} \\ b_{\mathbf{k}} \\ c_{\mathbf{k}} \end{pmatrix} \rightarrow \begin{pmatrix} a_{\mathbf{k}} \\ b_{\mathbf{k}} \\ c_{\mathbf{k}} \end{pmatrix} + \beta \begin{pmatrix} 0 & -\frac{\partial}{\partial k_2}\frac{1}{\sqrt{2}|\mathbf{k}|} & -\frac{\partial}{\partial k_2}\frac{1}{\sqrt{2}|\mathbf{k}|} \\ \frac{i}{\sqrt{2}|\mathbf{k}|}\frac{\partial}{\partial k_2} & 0 & -\frac{k_1}{4|\mathbf{k}|^3} \\ \frac{i}{\sqrt{2}|\mathbf{k}|}\frac{\partial}{\partial k_2} & \frac{k_1}{4|\mathbf{k}|^3} & 0 \end{pmatrix} \begin{pmatrix} a_{\mathbf{k}} \\ b_{\mathbf{k}} \\ c_{\mathbf{k}} \end{pmatrix}. \tag{2.41}
$$

Then the system in the leading order takes the form:

$$
\begin{pmatrix} \partial a_{\mathbf{k}}/\partial t \\ \partial b_{\mathbf{k}}/\partial t \\ \partial c_{\mathbf{k}}/\partial t \end{pmatrix} + \begin{pmatrix} i\Omega_{\mathbf{k}} & 0 & 0 \\ 0 & i\omega_{\mathbf{k}} & 0 \\ 0 & 0 & -i\omega_{-\mathbf{k}} \end{pmatrix} \begin{pmatrix} a_{\mathbf{k}} \\ b_{\mathbf{k}} \\ c_{\mathbf{k}} \end{pmatrix} +
$$

$$
+ \begin{pmatrix} \int (U_{klm}^{(0)}a_1 a_{\mathbf{m}} + U_{klm}^{(1)}a_1 b_{\mathbf{m}} + U_{klm}^{(2)}a_1 c_{\mathbf{m}})d\lambda \\ \int (V_{klm}^{(0)}a_1 b_{\mathbf{m}} + V_{klm}^{(1)}a_1 a_{\mathbf{m}} + V_{klm}^{(2)}a_1 c_{\mathbf{m}})d\lambda \\ \int (W_{klm}^{(0)}a_1 c_{\mathbf{m}} + W_{klm}^{(1)}a_1 a_{\mathbf{m}} + W_{klm}^{(2)}a_1 b_{\mathbf{m}})d\lambda \end{pmatrix} +
$$

$$
+ \begin{pmatrix} 0 \\ \int (V_{klm}^{(3)}b_1 b_{\mathbf{m}} + V_{klm}^{(4)}c_1 c_{\mathbf{m}} + V_{klm}^{(5)}b_1 c_{\mathbf{m}})d\lambda \\ \int (W_{klm}^{(3)}c_1 c_{\mathbf{m}} + W_{klm}^{(4)}b_1 b_{\mathbf{m}} + W_{klm}^{(5)}c_1 b_{\mathbf{m}})d\lambda \end{pmatrix} = 0, \tag{2.42}
$$

So Rossby waves split out, i. e. if $a_{\mathbf{k}} = 0$ at the initial moment, then $a_{\mathbf{k}} = 0$ for all times while these equation are applicable. As a result we obtain separate equations for short inertia-gravity waves

$$
\begin{pmatrix} \partial b_{\mathbf{k}}/\partial t \\ \partial c_{\mathbf{k}}/\partial t \end{pmatrix} + \begin{pmatrix} i\omega_{\mathbf{k}} & 0 \\ 0 & -i\omega_{-\mathbf{k}} \end{pmatrix} \begin{pmatrix} b_{\mathbf{k}} \\ c_{\mathbf{k}} \end{pmatrix} +
$$

$$
+ \begin{pmatrix} \int (V_{klm}^{(3)}b_1 b_{\mathbf{m}} + V_{klm}^{(4)}c_1 c_{\mathbf{m}} + V_{klm}^{(5)}b_1 c_{\mathbf{m}})d\lambda \\ \int (W_{klm}^{(3)}c_1 c_{\mathbf{m}} + W_{klm}^{(4)}b_1 b_{\mathbf{m}} + W_{klm}^{(5)}c_1 b_{\mathbf{m}})d\lambda \end{pmatrix} = 0. \tag{2.43}
$$

These equations are equivalent to the following Hamiltonian equations:

$$
\begin{pmatrix} \dot{\varphi} \\ \dot{z} \end{pmatrix} + \begin{pmatrix} 0 & 1 \\ -1 & 0 \end{pmatrix} \begin{pmatrix} \delta H/\delta\varphi \\ \delta H/\delta z \end{pmatrix} = 0, \tag{2.44}
$$

with Hamiltonian

$$
H = \int \left(\frac{1}{2}(1+z)(\varphi_x^2 + \varphi_y^2) + \frac{1}{2}z^2 + \beta\varphi\Delta^{-1}z_x \right) dxdy, \tag{2.45}
$$

in terms of two space-time variables $\phi(\mathbf{r}, t)$, $z(\mathbf{r}, t)$ with Fourier - transforms:

$$
\begin{pmatrix} \varphi_{\mathbf{k}} \\ z_{\mathbf{k}} \end{pmatrix} = \begin{pmatrix} \frac{i}{\sqrt{2}k} & -\frac{i}{\sqrt{2}k} \\ \frac{1}{\sqrt{2}} & \frac{1}{\sqrt{2}} \end{pmatrix} \begin{pmatrix} b_{\mathbf{k}} \\ c_{\mathbf{k}} \end{pmatrix}. \tag{2.46}
$$

For real initial variables we have $c_{\mathbf{k}} = b_{-\mathbf{k}}^*$ therefore $b_{\mathbf{k}}^*$ can be considered as an indepedent variable.

The equations have a stardard Hamiltonian form

$$\frac{\partial b_{\mathbf{k}}}{\partial t} + i\frac{\delta H}{\delta b_{\mathbf{k}}^*} = 0 \tag{2.47}$$

with the Hamiltonian $H = H_2 + H_3$:

$$H_2 = \int \omega_{\mathbf{k}} \mid b_{\mathbf{k}} \mid^2 d\mathbf{k}, \quad \omega_{\mathbf{k}} = k - \frac{\beta}{2}\frac{k_x}{k^2}, \quad k = \sqrt{k_x^2 + k_y^2}, \tag{2.48}$$

where the frequency $\omega_{\mathbf{k}}$ is positive for short waves,

$$H_3 = \frac{1}{2}\int V_{123}\,(b_1 b_2^* b_3^* + b_1^* b_2 b_3)\,\delta(\mathbf{k}_1 - \mathbf{k}_2 - \mathbf{k}_3)d\mathbf{k}_1 d\mathbf{k}_2 d\mathbf{k}_3 +$$

$$+ \frac{1}{3}\int U_{123}\,(b_1 b_2 b_3 + b_1^* b_2^* b_3^*)\,\delta(\mathbf{k}_1 + \mathbf{k}_2 + \mathbf{k}_3)d\mathbf{k}_1 d\mathbf{k}_2 d\mathbf{k}_3. \tag{2.49}$$

Here in the leading order in β

$$2U_{123} = V_{123} = \sqrt{18}\frac{k_1(\mathbf{k}_2, \mathbf{k}_3) + k_2(\mathbf{k}_3, \mathbf{k}_1) + k_3(\mathbf{k}_1, \mathbf{k}_2)}{\sqrt{k_1 k_2 k_3}}.$$

Therefore, the standard methods of weak turbulence may be applied to equatorial SIGW. The results will, however, depend on the decay or non-decay character of the dispersion law (2.48). The following analysis shows that the dispersion law *changes its type* depending on orientation of the wave-vectors. We consider the three- wave synchronism conditions

$$\mathbf{k} = \mathbf{k}_1 + \mathbf{k}_2, \quad \omega_{\mathbf{k}} = \omega_{\mathbf{k}_1} + \omega_{\mathbf{k}_2}, \tag{2.50}$$

and redirect the coordinate system along the vector $\mathbf{k}$. Then

$$k = k_1 \cos\theta_1 + k_2 \cos\theta_2, \quad 0 = k_1 \sin\theta_1 + k_2 \sin\theta_2, \tag{2.51}$$

$$k_1 + k_2 - k - \frac{\beta}{2}\left(\frac{\cos(\alpha + \theta_1)}{k_1} + \frac{\cos(\alpha + \theta_2)}{k_2} - \frac{\cos\alpha}{k}\right) = 0, \tag{2.52}$$

where α is the angle between $\mathbf{k}$ and the direction $\mathbf{e}_x$. We rewrite the frequency equation as

$$\left(2k_1 \sin^2\frac{\theta_1}{2} + 2k_2 \sin^2\frac{\theta_2}{2}\right) -$$

$$- \frac{\beta\cos\alpha}{2}\left(\frac{\cos\theta_1}{k_1} + \frac{\cos\theta_2}{k_2} - \frac{1}{k}\right) - \frac{\beta\sin\alpha}{2}\left(\frac{\sin\theta_1}{k_1} + \frac{\sin\theta_2}{k_2}\right) = 0, \tag{2.53}$$

The characteristic scales are $k_1 \sim k_2 \sim 1 \gg \beta > 0$, therefore angles are small $\theta_1, \theta_2 \ll 1$.

Thus two situations are possible. First, assume that the third bracket is small: $\cos\alpha \gg \theta_j \sin\alpha$. Then the first and the second ones are in balance $\theta_j^2 \sim \beta\cos\alpha$. From these relations it is easy to find an applicability condition

$$\frac{\cos\alpha}{\sin^2\alpha} \gg \beta. \tag{2.54}$$

Second, if the second bracket is small $\cos \alpha \ll \theta_j \sin \alpha$ then $\theta_j \sim \beta \sin \alpha$ and therefore

$$\frac{\cos \alpha}{\sin^2 \alpha} \ll \beta. \tag{2.55}$$

In the first case we have the following balance in the main order

$$2k_1 \sin^2 \frac{\theta_1}{2} + 2k_2 \sin^2 \frac{\theta_2}{2} = \frac{\beta \cos \alpha}{2} \left(\frac{1}{k_1} + \frac{1}{k_2} - \frac{1}{k} \right) \tag{2.56}$$

and since

$$\frac{1}{k_1} + \frac{1}{k_2} - \frac{1}{k_1 + k_2} = \frac{(k_1 + k_2/2)^2 + 3k_2^2/4}{k_1 k_2 (k_1 + k_2)} > 0$$

a solution exists only for $\cos \alpha > 0$!

In the second case we have

$$2k_1 \sin^2 \frac{\theta_1}{2} + 2k_2 \sin^2 \frac{\theta_2}{2} = \frac{\beta \sin \alpha}{2} \left(\frac{\sin \theta_1}{k_1} + \frac{\sin \theta_2}{k_2} \right) \tag{2.57}$$

and $\theta_1, \theta_2 \sim \beta \ll 1$. Existence of solution is independent of the sign of $\sin \alpha$ because there always exist a trivial solution $\theta_1 = \theta_2 = 0$.

Hence, *resonant triads exist 1) in a relatively wide segment around x -axis in the right half-plane in the* **k** *- space; 2) in two very narrow segments around the y - axis. Apart from the narrow regions around the y - axis, there are no resonant triads in the left half-plane in the* **k** *- space.*

The kinetic equations are, therefore, different in the regions of the phase-space with allowed and forbidden resonant triads. In the first case the standard three-wave kinetic equation applies.

$$\begin{aligned}
\frac{\partial N(\mathbf{k}, t)}{\partial t} &= \pi \int [\, |V_{k12}|^2 \, f_{k12} \delta(\mathbf{k} - \mathbf{k}_1 - \mathbf{k}_2) \delta(\omega_k - \omega_1 - \omega_2) \\
&\quad - |V_{12k}|^2 \, f_{12k} \delta(\mathbf{k}_1 - \mathbf{k}_2 - \mathbf{k}) \delta(\omega_1 - \omega_2 - \omega_k) \\
&\quad - |V_{2k1}|^2 \, f_{2k1} \delta(\mathbf{k}_2 - \mathbf{k}_k - \mathbf{k}_1) \delta(\omega_2 - \omega_k - \omega_1) \,] \, d\mathbf{k}_1 d\mathbf{k}_2,
\end{aligned} \tag{2.58}$$

where $f_{k12} = N_1 N_2 - N N_1 - N N_2$ and $N_j = N(\mathbf{k}_j, t)$.

Dispersion is weak and using techniques similar to those used to factorize the four-wave collision integral in the previous subsections, the collision integral in the case $\cos \alpha \gg \beta \sin \alpha$, may be factorized for $N(\mathbf{k}) = k^s$:

$$I(\mathbf{k}) = \pi k^r \int |V_{k12}|^2 \, f_{k12} \delta(\mathbf{k} - \mathbf{k}_1 - \mathbf{k}_2) \delta(k_1 \theta_1 + k_2 \theta_2) \tag{2.59}$$

$$\delta \left(k_1 \theta_1^2 / 2 + k_2 \theta_2^2 / 2 - \frac{\beta \cos \alpha}{2} \left(\frac{1}{k_1} + \frac{1}{k_2} - \frac{1}{k} \right) \right) \tag{2.60}$$

$$\left(k^{-r} - k_1^{-r} - k_2^{-r} \right) d\mathbf{k}_1 d\mathbf{k}_2. \tag{2.61}$$

Here $r = 2s + 7$ is a sum of powers coming from: f which give $2s$, the interaction coefficient which gives 3, the delta-functions which give $-1 + 0 + 1 = 0$, and Jacobians which give 4.

As a result for $r = -1$ we have a Kolmogorov spectrum with $s = -4$ ($N \propto k^{-4}$). The spectral energy density is $\varepsilon_k = \omega_k k N_k \propto k^{-2}$. An equilibrium solution corresponds to $s = -1$ and equipartition of energy.

Curiously, the case of almost vertical resonant triads $\cos \alpha \ll \beta \sin \alpha$ does not allow such treatment.

A four wave collision integral may be factorized as well in the non-decay regions of the phase - space giving:

$$I = k^r \int |T_{k123}|^2 \, \delta(k + k_1 - k_2 - k_3)$$

$$\delta \left(k_1 \theta_1^2/2 - k_2 \theta_2^2/2 - k_3 \theta_3^2/2 - \frac{\beta \cos \alpha}{2} \left(k^{-1} + k_1^{-1} - k_2^{-1} - k_3^{-1} \right) \right)$$

$$\left(N_k^{-1} + N_1^{-1} - N_2^{-1} - N_3^{-1} \right) \left(k^{-r} + k_1^{-r} - k_2^{-r} - k_3^{-r} \right) dk_1 dk_2 dk_3$$

with $r = 3s + 13$. We thus get two stationary solutions: one with a constant energy flux: $r = -1$ and $N \propto k^{-14/3}$, and one with a constant wave-action: $r = 0$ and $N \propto k^{-13/3}$.

2.5   Weak turbulence of the Rossby waves on the mid-latitude β - plane

By introducing a nonlinearity parameter ϵ in the quasi-geostrophic equation on the beta - plane (2.7), and assuming that it is small we get:

$$\partial_t \left(h - \nabla^2 h \right) - \beta \partial_x h - \epsilon \mathcal{J} \left(h, \nabla^2 h \right) = 0. \tag{2.62}$$

Introducing Gaussian statistics for an ensemble of linear Rossby waves a following kinetic equation for the spectral density $F_{\mathbf{k}}$ of the height field h may be easily established by taking the Fourier- transform of (2.62), multiplying by the complex conjugate of $h_{\mathbf{k}}$, averaging and using the expansion of h in ϵ:

$$\dot{F}_{\mathbf{k}} = 4\pi \int d\mathbf{k}_1 d\mathbf{k}_2 \, \delta(\mathbf{k} + \mathbf{k}_1 + \mathbf{k}_2) \delta(\omega(\mathbf{k}) + \omega(\mathbf{k}_1) + \omega(\mathbf{k}_2)) \cdot$$

$$\frac{D_{\mathbf{k}_1 \mathbf{k}_2}}{(k^2 + 1)(k_1^2 + 1)(k_2^2 + 1)} \left(D_{\mathbf{k}_1 \mathbf{k}_2} F_{\mathbf{k}_1} F_{\mathbf{k}_2} + D_{\mathbf{k} \mathbf{k}_2} F_{\mathbf{k}} F_{\mathbf{k}_1} + D_{\mathbf{k} \mathbf{k}_2} F_{\mathbf{k}} F_{\mathbf{k}_2} \right) \cdot$$

$$\tag{2.63}$$

Here

$$D_{\mathbf{k}_1 \mathbf{k}_2} = \frac{1}{4\pi} \mathbf{k}_1 \times \mathbf{k}_2 \left(\mathbf{k}_2^2 - \mathbf{k}_1^2 \right), \tag{2.64}$$

and $\omega = -\beta \frac{k_x}{k^2 + 1}$. The derivation is analogous to that for internal gravity waves in the stratified fluid given in the next Section and is not repeated here.

The specificity of Rossby waves is that equation (2.62) is Hamiltonian, but not canonical. This is reflected, in particular, in the fact that it possesses an *infinity of integrals of motion*. Indeed, rewriting (2.7) in the form:

$$\partial_t \left(h - \nabla^2 h \right) + \mathcal{J} \left(h, h - \nabla^2 h - \beta y \right) = 0, \tag{2.65}$$

one sees that any function of $h - \nabla^2 h - \beta y$ integrated over the domain occupied by the flow is conserved. The Hamiltonian structure of the flow is given by the following Lie - Poisson bracket defined for any pair A and B of functionals of the *quasi-geostrophic relative potential vorticity* $q_{QG} = h - \nabla^2 h$:

$$\{A[q], B[q]\} = - \int dx dy \, (q - \beta y) \mathcal{J} \left(\frac{\delta A}{\delta q}, \frac{\delta B}{\delta q} \right). \tag{2.66}$$

The Hamiltonian is quadratic: $H = \frac{1}{2} \int dx dy \, h(h - \nabla^2 h)$.

For configurations with *no closed lines of* $q_{QG} - \beta y$ (regions with closed contours correspond to *vortices* while *waves* have isolines of $q_{QG} - \beta y$ slightly deviating from the straight lines $y = const$) there exist a change of independent variables which allows to transform the non-canonical Poisson bracket (2.66) into the canonical one. (This transformation, in fact, straightens the isolines of $q_{QG} - \beta y$ in the wave case). The transformation adds nonlinear terms to the initial quadratic Hamiltonian. The standard Hamiltonian weak-turbulence approach may be then applied. It may be shown, however, that the resulting three-wave kinetic equation is identical to (2.63).

Following the same lines as in the subsequent section, the collision integral for Rossby waves may be represented as a sum of three integrals and, for short waves with $\omega = -\beta \frac{k_x}{|\mathbf{k}|^2}$, such that $k_x \ll k_y$, the integration domains may be transformed one into another using the integration variables $s_j = \frac{k_{jx}}{k_x}$ and $t_j = \frac{\omega(k_j)}{\omega(k)}$. Supposing a self-similar solution $F(k_x, k_y) \propto k_x^\alpha k_y^\beta$, the whole collision integral then is factorized and the following stationary solutions are obtained:

$$F_k \propto k_x^{-\frac{3}{2}}, \quad F_k \propto k_x^{-\frac{3}{2}} k_y^{-1}. \tag{2.67}$$

It may be checked by direct substitution into (2.63) that generalized thermodynamic equilibria of the form:

$$F(\mathbf{k}) = \frac{1}{a + b k^2} + \Phi(k_y) \delta(k_x), \tag{2.68}$$

where a, b arbitrary constants, and Φ - arbitrary function, annihilate the collision integral.

The first part of the solution corresponds to energy equipartition, while the second gives an equilibrium with arbitrary zonal current, which is a peculiar property of the Rossby waves dynamics

2.6   Historical remarks and bibliography

The RSW model is a standard tool in geophysical fluid dynamics, for separation in slow and fast components and their non-interaction cf, e.g. Zeitlin, Reznik and Ben Jelloul (2001) and references therein.

Factorization of the collision integral in the non-decay case close to the acoustic law follows a virtually unknown paper of Volotsky, Kats and Kontorovich (1980) which was published in a local journal.

Kinetic equation and weak turbulence spectra of short SIGW on the f - plane were obtained in Falkovich and Medvedev (1992) , following the ideas of Volotsky, Kats and Kontorovich (1980) .

For a recent review of equatorial waves in the framework of RSW model see, e.g. Le Sommer, Reznik and Zeitlin (2004) . The results on weak turbulence of short equatorial waves (Medvedev and Zeitlin, 2004) are new.

Weak turbulence of the Rossby waves has a long history. Kinetic equation was obtained by Kenyon (1964) and Longuet-Higgins and Gill (1967) . Equilibrium spectra were derived and investigated by Reznik and Soomere (1983) and Reznik (1984) . The canonical variables for the equation (2.7) were obtained by Zakharov and Piterbarg (1987) and used to construct the kinetic equation. Relation between canonical and non-canonical Hamiltonian structure for nonlinear Rossby waves was established by Zeitlin (1992) . Equivalence of thus obtained kinetic equation and the old one of was established by Monin and Piterbarg (1987) and the nonequilibrium Kolmogorov-type spectra were found following the method of Kuznetsov (1972) . This method is also used below for internal gravity waves in the stratified fluid. Stability of non-equilibrium power-law solutions of the kinetic equation for Rossby waves was studied by Balk and Nazarenko (1990) .

3 Weak turbulence of internal gravity waves: example of strongly anisotropic medium

The starting point is stratified ideal fluid equations in the Boussinesq approximation written for velocity $\mathbf{v} = (u, v, w)$ and buoyancy $\xi = -g\frac{\rho - \rho_0}{\rho_0}$. $\rho_0(x_3)$ is a background vertical density stratification, $N^2 = -g\frac{d\rho_0}{dx_3}/\rho_0$ is the square of the Brunt-Väisälä frequency which is supposed to be constant, g is acceleration due to gravity (ξ should be replaced by potential temperature in atmospheric applications)

$$
\begin{aligned}
\dot{\mathbf{v}} + \mathbf{v}\nabla\mathbf{v} + \nabla P - \xi\hat{\mathbf{x}}_3 &= 0 \\
\dot{\xi} + \mathbf{v}\nabla\xi + N^2 w &= 0 \\
\nabla \cdot \mathbf{v} &= 0.
\end{aligned}
\tag{3.1}
$$

3.1 Kinetic equation and energy spectra for plane-parallel internal gravity waves

We start by considering 2d situation with no dependence on one of the horizontal coordinates, x_2. In this case the Boussinesq equations for the (vertical) streamfunction ψ and buoyancy ξ (potential temperature in the atmospheric context) variables are

$$
\Delta\dot{\psi} + J(\Delta\psi, \psi) + \xi_{x_1} = 0, \qquad \dot{\xi} + J(\xi, \psi) - N^2\psi_{x_1} = 0,
\tag{3.2}
$$

where the subscripts denote partial derivatives with respect to the corresponding argument. The velocity components are

$$u = \frac{\partial \psi}{\partial x_3}, \qquad v = -\frac{\partial \psi}{\partial x_1}.$$

The approximation is valid for short waves such that wave- amplitude growth with altitude due to decreasing background density may be neglected.

We assume that the wave field is weakly nonlinear, introduce a non-dimensional small amplitude ϵ and expand ψ and ξ:

$$\psi = \epsilon \psi^{(0)} + \epsilon^2 \psi^{(1)} + \cdots, \qquad \xi = \epsilon \xi^{(0)} + \epsilon^2 \xi^{(1)} + \cdots.$$

In the first order we get

$$\dot{\psi}^{(0)} = -\Delta^{-1}\xi^{(0)}_{x_1}, \quad \dot{\xi}^{(0)} = N^2 \psi^{(0)}_{x_1}, \tag{3.3}$$

which gives the dispersion relation for internal gravity waves:

$$\omega^2 = N^2 \frac{p_H^2}{p^2}. \tag{3.4}$$

We chose to split waves in a standard way, according to their frequency sign, i.e. $\omega = \pm\Omega_p$ with $\Omega_p = N\frac{|p_1|}{p} = N\frac{\hat{p}_1}{p}$, $\mathbf{p} = (p_1, p_3)$ and a hat over any function of p means multiplication by $\mathrm{sign}[p_1]$.

Solution of (3.3) in Fourier-space may be written as

$$\psi^{(0)}_{\mathbf{p}} = a_{\mathbf{p}} + \bar{a}_{-\mathbf{p}}, \quad \xi^{(0)}_{\mathbf{p}} = -N\hat{p}\left(a_{\mathbf{p}} - \bar{a}_{-\mathbf{p}}\right). \tag{3.5}$$

where $a_{\mathbf{p}} = \varphi^+(\mathbf{p})\,e^{i\Omega_p t}$ and $\varphi^- = \overline{\varphi^+}$ (the overbar denotes complex conjugation). In principle, mean flow and buoyancy corrections proportional to $\delta(p_1)$ may be added. However, as total (spatially integrated) buoyancy and vorticity are exactly conserved we may exclude these corrections in what follows by limiting ourselves by the states with zero total buoyancy and vorticity.

The energy of the system is a sum of kinetic and available potential energies

$$E = \frac{1}{2}\int d\mathbf{x}\,(-\psi\Delta\psi + N^{-2}\xi^2) = \frac{1}{2}\int d\mathbf{p}\,(p^2\,\psi_{\mathbf{p}}\,\psi_{-\mathbf{p}} + N^{-2}\xi_{\mathbf{p}}\,\xi_{-\mathbf{p}})$$

and for linear waves equipartition of energy between kinetic and potential parts takes place:

$$E = \int d\mathbf{p}\,p^2(a_{\mathbf{p}}\,\bar{a}_{\mathbf{p}} + a_{-\mathbf{p}}\,\bar{a}_{-\mathbf{p}}). \tag{3.6}$$

We calculate then the first nonlinear corrections by using the retarded Green's function (an infinitesimal damping of the linear waves δ is introduced shifting the poles of this function which is a solution of (3.33) upward from the real ω-axis). As a result, in the second order we get

$$\psi^{(1)}(\mathbf{p}, t)_{p_1 \neq 0} = -i \int d\mathbf{k} \, d\mathbf{l} \; \delta(\mathbf{k} + \mathbf{l} - \mathbf{p}) \sum_{\pm} A^{\pm}_{kl,p}$$

$$[\frac{a_{\mathbf{k}} a_{\mathbf{l}}}{\Omega_k + \Omega_l \mp \Omega_p - i\delta} + \frac{a_{\mathbf{k}} \overline{a}_{-\mathbf{l}}}{\Omega_k - \Omega_l \mp \Omega_p - i\delta}$$

$$- \frac{\overline{a}_{-\mathbf{k}} a_{\mathbf{l}}}{\Omega_k - \Omega_l \mp \Omega_p + i\delta} - \frac{\overline{a}_{-\mathbf{k}} \overline{a}_{-\mathbf{l}}}{\Omega_k + \Omega_l \mp \Omega_p + i\delta}], \qquad (3.7)$$

$$\xi^{(1)}(\mathbf{p}, t)_{p_1 \neq 0} = iN\hat{p} \int d\mathbf{k} \, d\mathbf{l} \; \delta(\mathbf{k} + \mathbf{l} - \mathbf{p}) \sum_{\pm} (\pm) A^{\pm}_{kl,p}$$

$$[\frac{a_{\mathbf{k}} a_{\mathbf{l}}}{\Omega_k + \Omega_l \mp \Omega_p - i\delta} + \frac{a_{\mathbf{k}} \overline{a}_{-\mathbf{l}}}{\Omega_k - \Omega_l \mp \Omega_p - i\delta}$$

$$+ \frac{\overline{a}_{-\mathbf{k}} a_{\mathbf{l}}}{\Omega_k - \Omega_l \mp \Omega_p + i\delta} + \frac{\overline{a}_{-\mathbf{k}} \overline{a}_{-\mathbf{l}}}{\Omega_k + \Omega_l \mp \Omega_p + i\delta}], \qquad (3.8)$$

where the interaction coefficients are defined as

$$A^{\pm}_{kl,p} = \mathbf{k} \times \mathbf{l} \, \frac{k^2 \pm \hat{p}\hat{k}}{2p^2}, \qquad \mathbf{k} \times \mathbf{l} \equiv k_1 l_3 - k_3 l_1. \qquad (3.9)$$

We next make a standard random phase approximation by supposing that due to the presence of a large amount of weakly interacting waves the complex wave amplitudes $a_{\mathbf{k}}$ assume Gaussian statistics

$$\begin{aligned}
< a_{\mathbf{k}} \, a_{\mathbf{l}} > &= 0, \\
< a_{\mathbf{k}} \, \overline{a}_{\mathbf{l}} > &= N_{\mathbf{k}} \, \delta(\mathbf{k} - \mathbf{l}), \\
< a_{\mathbf{k}} \, a_{\mathbf{l}} \, \overline{a}_{\mathbf{m}} \, \overline{a}_{\mathbf{n}} > &= N_{\mathbf{k}} \, N_{\mathbf{l}}(\delta(\mathbf{k} - \mathbf{m})\delta(\mathbf{l} - \mathbf{n}) + \delta(\mathbf{k} - \mathbf{n})\delta(\mathbf{l} - \mathbf{m})), \\
N_{\mathbf{k}} &= N_{-\mathbf{k}}.
\end{aligned} \qquad (3.10)$$

We denote abusively the quadratic mean N_k even though this is not a function of the modulus of $\mathbf{k}$. From (3.2) we obtain the following (exact) equations for averages of the Fourier components of the wave field:

$$\begin{aligned}
p^2 < \dot{\psi}_{\mathbf{p}} \, \psi_{\mathbf{q}} > &= \int d\mathbf{k} \, d\mathbf{l} \, V_{kl} \; < \psi_{\mathbf{k}} \, \psi_{\mathbf{l}} \, \psi_{\mathbf{q}} > \delta(\mathbf{k} + \mathbf{l} - \mathbf{p}) \\
&\quad - ip_1 < \xi_{\mathbf{p}} \, \psi_{\mathbf{q}} >, \\
< \dot{\xi}_{\mathbf{p}} \, \xi_{\mathbf{q}} > &= - \int d\mathbf{k} \, d\mathbf{l} \, W_{kl} \; < \xi_{\mathbf{k}} \, \psi_{\mathbf{l}} \, \xi_{\mathbf{q}} > \delta(\mathbf{k} + \mathbf{l} - \mathbf{p}) \\
&\quad - ip_1 N^2 < \psi_{\mathbf{p}} \, \xi_{\mathbf{q}} >,
\end{aligned} \qquad (3.11)$$

with $V_{kl} = k^2 \mathbf{k} \times \mathbf{l}$, $W_{kl} = -\mathbf{k} \times \mathbf{l}$.

By adding these two equations and their complex conjugates, we get the following kinetic equation for the spectral density of energy $\epsilon_{\mathbf{p}} = p^2(a_{\mathbf{p}} \bar{a}_{\mathbf{p}} + a_{-\mathbf{p}} \bar{a}_{-\mathbf{p}})$

$$< \dot{\epsilon}_{\mathbf{p}} >= \int d\mathbf{k}\, d\mathbf{l}\; \delta(\mathbf{k}+\mathbf{l}-\mathbf{p}) V_{kl} [< \psi_{\mathbf{k}} \psi_{\mathbf{l}} \psi_{-\mathbf{p}} > + < \psi_{-\mathbf{k}} \psi_{-\mathbf{l}} \psi_{\mathbf{p}} >]$$
$$- N^{-2} \int d\mathbf{k}\, d\mathbf{l}\; \delta(\mathbf{k}+\mathbf{l}-\mathbf{p}) W_{kl}\, [< \xi_{\mathbf{k}} \psi_{\mathbf{l}} \xi_{-\mathbf{p}} > + < \xi_{-\mathbf{k}} \psi_{-\mathbf{l}} \xi_{\mathbf{p}} >]$$

$$(3.12)$$

and calculate the r.h.s. of this equation perturbatively using (3.39) and Sokhotsky formula

$$\frac{1}{A \pm i\delta} = P\frac{1}{A} \mp i\pi\delta(A). \tag{3.13}$$

Non-resonant contributions to (3.12) cancel out pairwise and only the resonant terms remain. Using (3.7) , (3.8) and (3.13), we get,

$$< \dot{\epsilon}_{\mathbf{p}} >= 2\pi p^2 \int d\mathbf{k}\, d\mathbf{l}\; \delta(\mathbf{k}+\mathbf{l}-\mathbf{p})\{$$
$$A^+_{\{kl\},p}\, [A^+_{\{kl\},p} N_k N_l - A^+_{\{lp\},k} N_p N_l - A^+_{\{kp\},l} N_p N_k]\, \delta(\Omega_k + \Omega_l - \Omega_p)$$
$$+ A^{\star}_{kl,p}\, [A^{\star}_{kl,p} N_k N_l - A^-_{\{kp\},l} N_p N_k - A^{\star}_{pl,k} N_p N_l]\, \delta(\Omega_l + \Omega_p - \Omega_k)$$
$$+ A^{\star}_{lk,p}\, [A^{\star}_{lk,p} N_k N_l - A^-_{\{lp\},k} N_p N_l - A^{\star}_{pk,l} N_p N_k]\, \delta(\Omega_k + \Omega_p - \Omega_l)\}, \quad (3.14)$$

where we defined $A^{\pm}_{\{kl\},p} = A^{\pm}_{kl,p} + A^{\pm}_{kl,p}$ and $A^{\star}_{kl,p} = A^+_{kl,p} + A^-_{lk,p}$.

This is a typical kinetic equation with two δ-functions assuring that the collision process conserves energy and momentum. However, some transformations are necessary in order to cast it to the standard form. A change of variables should be made in order to have a standard form *frequency × wavenumber density* for $\epsilon_{\mathbf{p}}$

$$N_p \rightarrow N\frac{\hat{p}_1}{p^3} N_p \tag{3.15}$$

so that $\epsilon_{\mathbf{p}} = 2\Omega_p N_p$, N_p is the wavenumber density and is an adiabatic invariant in wave dynamics. We correspondingly modify the interaction coefficients

$$\mathcal{A}^{\pm}_{kl,p} = A^{\pm}_{\{kl\},p}(\frac{p}{kl})^{\frac{3}{2}}, \quad \mathcal{A}^{\star}_{kl,p} = A^{\star}_{kl,p}(\frac{p}{kl})^{\frac{3}{2}} \tag{3.16}$$

and get

$$< \dot{N}_{\mathbf{p}} >\sim \pi N \int d\mathbf{k}\, d\mathbf{l}\; \delta(\mathbf{k}+\mathbf{l}-\mathbf{p})\{$$
$$\mathcal{A}^+_{kl,p}\, \left[|\frac{k_1 l_1}{q_1}|\mathcal{A}^+_{kl,p} N_k N_l - |l_1|\mathcal{A}^+_{lp,k} N_p N_l - |k_1|\mathcal{A}^+_{kp,l} N_p N_k\right]\, \delta(\Omega_k + \Omega_l - \Omega_p) +$$
$$\mathcal{A}^{\star}_{kl,p}\, \left[|\frac{k_1 l_1}{q_1}|\mathcal{A}^{\star}_{kl,p} N_k N_l - |k_1|\mathcal{A}^-_{kp,l} N_p N_k - |l_1|\mathcal{A}^{\star}_{pl,k} N_p N_l\right]\, \delta(\Omega_l + \Omega_p - \Omega_k) +$$
$$\mathcal{A}^{\star}_{lk,p}\, \left[|\frac{k_1 l_1}{q_1}|\mathcal{A}^{\star}_{lk,p} N_k N_l - |l_1|\mathcal{A}^-_{lp,k} N_p N_l - |k_1|\mathcal{A}^{\star}_{pk,l} N_p N_k\right]\, \delta(\Omega_k + \Omega_p - \Omega_l)\}. \quad (3.17)$$

Taking into account the symmetry properties and the fact that the above expressions are to be calculated on the resonant surfaces defined by the arguments of the δ-functions, we have

$$|\frac{k_1\, l_1}{q_1}|\mathcal{A}^+_{kl,p} = |l_1|\mathcal{A}^+_{lp,k} = |k_1|\mathcal{A}^+_{kp,l}, |\frac{k_1\, l_1}{q_1}|\mathcal{A}^\star_{kl,p} = |l_1|\mathcal{A}^\star_{pl,k} = -|k_1|\mathcal{A}^-_{kp,l} \qquad (3.18)$$

and the equation (3.17) becomes

$$\dot{N}_p \ \sim \ \pi N \int d\mathbf{k}\, d\mathbf{l}\ \delta(\mathbf{k}+\mathbf{l}-\mathbf{p})\{$$
$$Z^+_{kl,p}\ (N_k N_l - N_p N_k - N_p N_l)\ \delta(\Omega_k + \Omega_l - \Omega_p) +$$
$$Z^-_{kp,l}\ (N_k N_l - N_p N_l + N_p N_k)\ \delta(\Omega_l + \Omega_p - \Omega_k) +$$
$$Z^-_{lp,k}\ (N_k N_l - N_p N_k + N_p N_l)\ \delta(\Omega_k + \Omega_p - \Omega_l)\} \qquad (3.19)$$

with

$$Z^\pm_{kl,p} = \frac{1}{4}\frac{\hat{k}_1\, \hat{l}_1\, \hat{p}_1}{klp}(\hat{k}+\hat{l}\pm\hat{p})^2\ (\frac{\hat{k}_3}{k} \mp \frac{\hat{l}_3}{l} \mp \frac{\hat{p}_3}{p})^2 . \qquad (3.20)$$

The sign in the first bracket of this expression corresponds to the superscript, the sign of the second bracket should be taken as that of the subsequent δ-function in (3.19).

Each integrand will be split in three to reduce the integration domain to $k_1 \geq 0$ and $l_1 \geq 0$ (we suppose then that $p_1 \geq 0$ as well):

$$\dot{N}_p \ \sim \ \pi N \int d\mathbf{k}\, d\mathbf{l}\ \delta(\mathbf{k}+\mathbf{l}-\mathbf{p})\{$$
$$Z^+_{kl,p}\ (N_k N_l - N_p N_k - N_p N_l)\ \delta(\Omega_k + \Omega_l - \Omega_p) +$$
$$Z^-_{kp,l}\ (N_k N_l - N_p N_l + N_p N_k)\ \delta(\Omega_l + \Omega_p - \Omega_k) +$$
$$Z^-_{lp,k}\ (N_k N_l - N_p N_k + N_p N_l)\ \delta(\Omega_k + \Omega_p - \Omega_l)\}$$
$$+ 2\pi N \int d\mathbf{k}\, d\mathbf{l}\ \delta(\mathbf{k}+\mathbf{p}-\mathbf{l})\{$$
$$Z^-_{lp,k}\ (N_k N_l - N_p N_k - N_p N_l)\ \delta(\Omega_k + \Omega_l - \Omega_p) +$$
$$Z^-_{lk,p}\ (N_k N_l - N_p N_l + N_p N_k)\ \delta(\Omega_l + \Omega_p - \Omega_k) +$$
$$Z^+_{lp,k}\ (N_k N_l - N_p N_k + N_p N_l)\ \delta(\Omega_k + \Omega_p - \Omega_l)\}. \qquad (3.21)$$

Due to the complex form of the function Z in (3.20) it is impossible to find exact solutions of (3.21) without making additional assumptions. We will try to find an analytic power law solution by limiting ourselves by waves with almost vertical phase propagation : $k_3 \gg k_1$. In this case $Z^\pm_{kl,p}$ greatly simplifies:

$$Z^\pm_{kl,p} = \frac{1}{4}\frac{|k_1\, l_1\, p_1|}{|k_3\, l_3\, p_3|}(\widehat{|k_3|} + \widehat{|l_3|} \pm \widehat{|p_3|})^2. \qquad (3.22)$$

We are then looking for spectra averaged in the vertical direction, i.e. our spectral density N_p will depend only on $|p_3|$. This allows us to reduce all integrations to the domain k_3, $l_3 \geq 0$ for $p_3 \geq 0$. In this way, each integral splits in four and for each sub-integral we may transform the integration variables according to

$$k_1 = s\,p_1, \qquad k_3 = \frac{s}{t}\,p_3, \qquad l_1 = u\,p_1, \qquad l_3 = \frac{u}{v}\,p_3. \qquad (3.23)$$

$(s > 0, t > 0, u > 0, v > 0)$ and look for a scaling solution

$$N_k \sim s^{\alpha} t^{\beta} N_p, \qquad N_l \sim u^{\alpha} v^{\beta} N_p, \qquad N_k \sim k_1^{\alpha} \Omega_k^{\beta}, \qquad N_k \sim k_1^{\alpha+\beta} k_3^{-\beta}. \qquad (3.24)$$

Performing frequency and the horizontal wavevector integration by using the corresponding δ-functions and making, when necessary a change of variables $s \to s/(1-s), t \to t/(1-t)$ in order to have a common bounded integration domain $((0,1),(0,1))$ we arrive to the expression (up to a constant) containing two different integrals:

$$
\begin{aligned}
\dot{N}_p \;\sim\; & \int_0^1 ds \int_0^1 dt\; \delta(1 \pm \frac{s}{t} \pm \frac{1-s}{1-t}) U(s,t) \times \\
& [s^{\alpha} t^{\beta}(1-s)^{\alpha}(1-t)^{\beta} - s^{\alpha} t^{\beta} - (1-s)^{\alpha}(1-t)^{\beta}] \times \\
& [1 - (1-s)^{-2\alpha-6}(1-t)^{-2\beta+2} - s^{-2\alpha-6} t^{-2\beta+2}] \\
+\;\; & 2 \int_0^1 ds \int_0^1 dt\; \delta(\frac{1}{t} \pm s \pm \frac{1-s}{1-t}) V(s,t) \times \\
& [s^{\alpha} t^{\beta} + s^{\alpha}(1-s)^{\alpha}(1-t)^{\beta} - t^{\beta}(1-s)^{\alpha}(1-t)^{\beta}] \times \\
& [t^{-2\beta+2} - s^{-2\alpha-6} + (1-s)^{-2\alpha-6}(1-t)^{-2\beta+2}]. \qquad (3.25)
\end{aligned}
$$

Here $U(s,t) = \frac{s(1-s)}{t(1-t)}(1 + \frac{s}{t} + \frac{1-s}{1-t})^2$ and $V(s,t) = \frac{s(1-s)}{t(1-t)}(\frac{1}{t} + s - \frac{1-s}{1-t})^2$. Note a difference with similar expression for *unidirectional* Rossby waves, where a single integral arises.

The solutions are a statistical equilibrium spectrum

$$N_k = 1/\Omega_k, \qquad \epsilon_k = const, \qquad (3.26)$$

and a single non-equilibrium spectrum

$$N_k \sim k_1^{-\frac{5}{2}} k_3^{-\frac{1}{2}}, \qquad \epsilon_k \sim k_1^{-\frac{3}{2}} k_3^{-\frac{3}{2}} \qquad \epsilon_k \sim \Omega_k^{-\frac{3}{2}} k_3^{-3}. \qquad (3.27)$$

The first one is energy equipartition, i.e. the Rayleigh-Jeans spectrum while the second one is a Kolmogorov-like spectrum corresponding to a constant energy flux through the spectrum. Indeed, let P be an energy flux. Then by dimensional reasons a spectrum constructed from P and the square of the Brunt-Väisälä frequency must be proportional to the wavenumber in the -3 power which corresponds to (3.27)

$$\epsilon_k = C_2\, N^{\frac{1}{2}} P^{\frac{1}{2}} k_1^{-\frac{3}{2}} k_3^{-\frac{3}{2}} \qquad (3.28)$$

3.2 Kinetic equation and horizontally isotropic stationary energy spectra for 3D internal gravity waves in the absence of mean potential vorticity

In the 3d case it is convenient to use the Craya-Herring decomposition for velocity:

$$\mathbf{v} = \nabla_H \times \psi\,\hat{x}_3 + \nabla_H \phi + w\,\hat{x}_3 \qquad (3.29)$$

which in the linear approximations gives a separation of the flow in a vortical (ψ) and wave (ϕ, w) parts.

$$
\begin{aligned}
\Delta_H \phi + \partial_3 w &= 0, \\
\Delta P - \partial_3 \xi + \nabla \cdot (\mathbf{v} \cdot \nabla \mathbf{v}) &= 0, \\
\dot{w} + \mathbf{v} \cdot \nabla w + \partial_3 P - \xi &= 0, \\
\dot{\xi} + \mathbf{v} \cdot \nabla \xi + N^2 w &= 0,
\end{aligned}
\tag{3.30}
$$

$$
-\Delta_H \dot{\psi} - J(\Delta_H \psi, \psi) - \nabla_H \cdot (\nabla_H \phi \Delta_H \psi) - w \partial_3 \Delta_H \psi \\
- \nabla_H w \cdot \nabla_H \partial_3 \psi + J(w, \partial_3 \phi) = 0, \tag{3.31}
$$

where the subscripts H and 3, denote the horizontal and vertical parts, respectively and

$$
\mathbf{v} \cdot \nabla \ldots = \nabla_H \phi \cdot \nabla \ldots + w \partial_3 + J(\ldots, \psi). \tag{3.32}
$$

By introducing Fourier transforms $\psi_\mathbf{p}$, $\xi_\mathbf{p}$, $\phi_\mathbf{p}$ and $w_\mathbf{p}$ and linearizing the system (3.1) one gets internal gravity waves solutions wi th a dispersion relation

$$
\omega^2 = N^2 \frac{\mathbf{p}_H^2}{\mathbf{p}^2} \tag{3.33}
$$

where ω is the wave frequency, $\mathbf{p} = (p_1, p_2, p_3)$ is the wavenumber, $p \equiv |\mathbf{p}|$ and $\mathbf{p}_H = (p_1, p_2)$ is its projection on the horizonta l plane. Note that $\psi_\mathbf{p}$ represents a zero mode of the linear system, i.e. is time-independent in the first approximation, and that mean flow/stratification c orrections proportional to $\delta(p_H)$ must be added to get a general solution of the linear system. We shall discard this contributions altogether by the reasons explained below. Note also that the wave part of $\phi_\mathbf{p}$ may be always eliminated in favor of $w_\mathbf{p}$.

By introducing wave amplitudes $b_\mathbf{p} = e^{i\Omega_p t} \chi_\mathbf{p}^+$ corresponding to the two branches of the dispersion equation (3.33), where Ω_p is the positive-sign root of (3.33), we get

$$
w_\mathbf{p} = b_\mathbf{p} + \bar{b}_{-\mathbf{p}}, \quad \xi_\mathbf{p} = iN \frac{p}{p_H} (b_\mathbf{p} - \bar{b}_{-\mathbf{p}}) \tag{3.34}
$$

The energy of the system in the absence of vortical part and mean buoyancy is

$$
\begin{aligned}
E &= \frac{1}{2} \int d\mathbf{x} \left(-\psi \Delta_H \psi - \phi \Delta_H \phi + w^2 + N^{-2} \xi^2 \right) \\
&= \frac{1}{2} \int d\mathbf{p} \left(p_H^2 \phi_\mathbf{p} \phi_{-\mathbf{p}} + w_\mathbf{p} w_{-\mathbf{p}} + N^{-2} \xi_\mathbf{p} \xi_{-\mathbf{p}} \right) \\
&= \frac{1}{2} \int d\mathbf{p} \left(2 \frac{p^2}{p_H^2} (b_\mathbf{p} \bar{b}_\mathbf{p} + b_{-\mathbf{p}} \bar{b}_{-\mathbf{p}}) \right)
\end{aligned}
\tag{3.35}
$$

We limit ourselves by a study of purely wave regimes, thus, excluding completely the vortical mode and mean flow/stratification corrections. A physical basis for this separation is the Lagrangian (pointwise) conservation of the potential vorticity $q = (\nabla \times \mathbf{v})(\nabla \xi + N^2 \hat{\mathbf{x}}_3)$. For a single harmonic wave potential vorticity is identically zero. This is not true anymore for a superposition of waves and, in fact, waves do produce vorticity and mean bouyancy already in the first order of perturbation theory. However, for a Gaussian ensemble of waves (see below) the statistically averaged potential vorticity is still zero which is not the case in the present of vortical modes or mean buoyancy corrections. We, thus, have a separation criterion and by limiting ourselves by zero mean potential vorticity configurations we discard all non-wave corrections.

The first order corrections, therefore, are

$$w^{(1)}(\mathbf{p}, t) = -\int d\mathbf{k}\, d\mathbf{l}\, \delta(\mathbf{k} + \mathbf{l} - \mathbf{p}) \sum_{\pm} B^{\pm}_{kl,p}$$

$$\left[\frac{b_{\mathbf{k}} b_{\mathbf{l}}}{\Omega_k + \Omega_l \mp \Omega_p - i\delta} + \frac{b_{\mathbf{k}} \bar{b}_{-\mathbf{l}}}{\Omega_k - \Omega_l \mp \Omega_p - i\delta} \right.$$

$$\left. - \frac{\bar{b}_{-\mathbf{k}} b_{\mathbf{l}}}{\Omega_k - \Omega_l \mp \Omega_p + i\delta} - \frac{\bar{b}_{-\mathbf{k}} \bar{b}_{-\mathbf{l}}}{\Omega_k + \Omega_l \mp \Omega_p + i\delta} \right] \qquad (3.36)$$

$$\xi^{(1)}(\mathbf{p}, t) = -iN \frac{p}{p_H} \int d\mathbf{k}\, d\mathbf{l}\, \delta(\mathbf{k} + \mathbf{l} - \mathbf{p}) \sum_{\pm} (\pm) B^{\pm}_{kl,p}$$

$$\left[\frac{b_{\mathbf{k}} b_{\mathbf{l}}}{\Omega_k + \Omega_l \mp \Omega_p - i\delta} + \frac{b_{\mathbf{k}} \bar{b}_{-\mathbf{l}}}{\Omega_k - \Omega_l \mp \Omega_p - i\delta} \right.$$

$$\left. + \frac{\bar{b}_{-\mathbf{k}} b_{\mathbf{l}}}{\Omega_k - \Omega_l \mp \Omega_p + i\delta} + \frac{\bar{b}_{-\mathbf{k}} \bar{b}_{-\mathbf{l}}}{\Omega_k + \Omega_l \mp \Omega_p + i\delta} \right] \qquad (3.37)$$

where

$$B^{\pm}_{kl,p} = (p_H^2 + p_3 l_3 \frac{\mathbf{p} \cdot \mathbf{l}}{l_H^2} \pm p l \frac{p_H}{l_H}) \frac{H_{kl}}{2p^2} \,, \quad H_{kl} = \frac{\mathbf{k} \cdot \mathbf{l}}{k_H^2} k_3 - l_3. \qquad (3.38)$$

We next make the random phase approximation and suppose that due to the presence of a large amount of weakly interacting waves the complex wave amplitudes $b_{\mathbf{k}}$ assume Gaussian statistics

$$\begin{aligned} <b_{\mathbf{k}} \bar{b}_{\mathbf{l}}> &= N_{\mathbf{k}}\, \delta(\mathbf{k} - \mathbf{l})\,, \\ <b_{\mathbf{k}} b_{\mathbf{l}} \bar{b}_{\mathbf{m}} \bar{b}_{\mathbf{n}}> &= N_{\mathbf{k}}\, N_{\mathbf{l}}(\delta(\mathbf{k} - \mathbf{m})\delta(\mathbf{l} - \mathbf{n}) + \delta(\mathbf{k} - \mathbf{n})\delta(\mathbf{l} - \mathbf{m}))\,, \end{aligned}$$

$$(3.39)$$

where the wavenumber density $N_{\mathbf{k}} = N_{-\mathbf{k}}$ is a real-valued function and we again denote abusively the quadratic mean by N_k.

From (3.1) we obtain the following (exact) equations for averages of the Fourier components of the wave field:

$$< \dot{w}_\mathbf{p}\, w_\mathbf{q} > \;=\; \frac{p_H^2}{p^2} < \xi_\mathbf{p}\, w_\mathbf{q} >$$

$$- \frac{i\, p_H^2}{p^2} \int dk\, dl\; \delta(\mathbf{k}+\mathbf{l}-\mathbf{p})\; I_{kl,p} < w_k\, w_l\, w_\mathbf{q} >,$$

$$< \dot{\xi}_\mathbf{p}\, \xi_\mathbf{q} > \;=\; -N^2 < w_\mathbf{p}\, \xi_\mathbf{q} >$$

$$- i \int dk\, dl\; \delta(\mathbf{k}+\mathbf{l}-\mathbf{p})\; H_{kl} < w_k\, \xi_l\, \xi_\mathbf{q} > . \tag{3.40}$$

where $I_{kl,p} = (1 + \frac{p_3 l_3}{l_H^2}\frac{\mathbf{p}\cdot\mathbf{l}}{p_H^2})\, H_{kl}$.

By adding these two equations and their complex conjugates, we get the following kinetic equation for the energy spectral density $\epsilon_\mathbf{p} = \frac{p^2}{p_H^2}(b_\mathbf{p}\,\bar{b}_\mathbf{p} + b_{-\mathbf{p}}\,\bar{b}_{-\mathbf{p}})$

$$< \dot{\epsilon}_p > = \frac{i}{2} \int dk\, dl\; \delta(\mathbf{k}+\mathbf{l}-\mathbf{p})\; I_{kl,p}\; [< w_{-k}\, w_{-l}\, w_\mathbf{p} > - < w_k\, w_l\, w_{-\mathbf{p}} >]$$

$$+ \frac{i}{2N^2} \int dk\, dl\; \delta(\mathbf{k}+\mathbf{l}-\mathbf{p})\; H_{kl}\; [< w_{-k}\, \xi_{-l}\, \xi_\mathbf{p} > - < w_k\, \xi_l\, \xi_{-\mathbf{p}} >] \tag{3.41}$$

and calculate the r.h.s. of this equation perturbatively using (3.39) and Sokhotsky formulas.

We get

$$< \dot{\epsilon}_\mathbf{p} > \;\sim\; 2\pi \frac{p^2}{p_H^2} \int dk\, dl\; \delta(\mathbf{k}+\mathbf{l}-\mathbf{p})\{$$

$$B_{\{kl\},p}^+ \quad [B_{\{kl\},p}^+ N_k N_l - D_{lp,k}^- N_p N_l - D_{kp,l}^- N_p N_k]\; \delta(\Omega_k + \Omega_l - \Omega_p) +$$

$$D_{lk,p}^+ \quad [D_{pk,l}^- N_p N_k + C_{pl,k}^+ N_p N_l + D_{lk,p}^+ N_k N_l]\; \delta(\Omega_l + \Omega_p - \Omega_k) +$$

$$D_{kl,p}^+ \quad [D_{pl,k}^- N_p N_l + C_{pk,l}^+ N_p N_k + D_{kl,p}^+ N_k N_l]\; \delta(\Omega_k + \Omega_p - \Omega_l)\}$$

$$\tag{3.42}$$

where $C_{kl,p}^\pm = B_{kl,p}^\pm - B_{lk,p}^\pm$, $D_{kl,p}^\pm = B_{kl,p}^+ \pm B_{lk,p}^-$.

This is a typical kinetic equation with two δ-functions assuring that the collision process conserves energy and momentum. However, some transformations are necessary in order to cast it to the standard form. A change of variables should be made in order to have a standard *frequency × wavenumber density* form for $\epsilon_\mathbf{p}$:

$$N_p \to N \frac{p_H^3}{p^3} N_p$$

thus giving $\epsilon_p = 2\Omega_p N_p$ in terms of new N_p which is usually interpreted as "quasiparticle" density. We correspondingly modify the interaction coefficients

$$\mathcal{B}_{kl,p} = B_{\{kl\},p}^+ (\frac{p}{kl})^{\frac{3}{2}}, \qquad \mathcal{C}_{kl,p} = C_{kl,p}^+ (\frac{p}{kl})^{\frac{3}{2}} \qquad \mathcal{D}_{kl,p}^\pm = D_{kl,p}^\pm (\frac{p}{kl})^{\frac{3}{2}} \tag{3.43}$$

and finally get:

$$
< \dot{N}_{\mathbf{p}} > \sim \pi N \int d\mathbf{k}\, d\mathbf{l}\, \delta(\mathbf{k}+\mathbf{l}-\mathbf{p})\{
$$

$$
\mathcal{B}_{\{kl\},p} \left[\frac{k_H^3 l_H^3}{p_H^3} \mathcal{B}_{\{kl\},p} N_k N_l - k_H^3 \mathcal{D}_{lp,k}^- N_p N_l - l_H^3 \mathcal{D}_{kp,l}^- N_p N_k \right] \delta(\Omega_k + \Omega_l - \Omega_p) +
$$

$$
\mathcal{D}_{lk,p}^+ \left[k_H^3 \mathcal{D}_{pk,l}^- N_p N_k + l_H^3 \mathcal{C}_{pl,k}^+ N_p N_l + \frac{k_H^3 l_H^3}{p_H^3} \mathcal{D}_{lk,p}^+ N_k N_l \right] \delta(\Omega_l + \Omega_p - \Omega_k) +
$$

$$
\mathcal{D}_{kl,p}^+ \left[l_H^3 \mathcal{D}_{pl,k}^- N_p N_l + k_H^3 \mathcal{C}_{pk,l}^+ N_p N_k + \frac{k_H^3 l_H^3}{p_H^3} \mathcal{D}_{kl,p}^+ N_k N_l \right] \delta(\Omega_k + \Omega_p - \Omega_l) \} \quad (3.44)
$$

we limit ourselves by systems of waves (wave packets) propagating almost vertically, i.e. $k_3 \gg k_H$ and, thus, having low frequencies. In this case we get

$$
\frac{k_H^3 l_H^3}{p_H^3} \mathcal{B}_{kl,p} = k_H^3 \mathcal{D}_{kp,l}^- = l_H^3 \mathcal{D}_{lp,k}^-, \quad l_H^3 \mathcal{C}_{pl,k} = -k_H^3 \mathcal{D}_{pk,l}^- = -\frac{k_H^3 l_H^3}{p_H^3} \mathcal{D}_{lk,p}^+ \quad (3.45)
$$

and the collision integral in the r.h.s. of (3.44) greatly simplifies

$$
< \dot{N}_{\mathbf{p}} > \quad \sim \quad \pi N \int d\mathbf{k}\, d\mathbf{l}\, \delta(\mathbf{k}+\mathbf{l}-\mathbf{p})\{
$$

$$
X_{kl,p}[N_k N_l - N_p N_k - N_p N_l]\, \delta(\Omega_k + \Omega_l - \Omega_p) +
$$

$$
Y_{kl,p}^+[N_k N_l + N_p N_k - N_p N_l]\, \delta(\Omega_l + \Omega_p - \Omega_k) +
$$

$$
Y_{lk,p}^+[N_k N_l + N_p N_l - N_p N_k]\, \delta(\Omega_k + \Omega_p - \Omega_l) \}
$$

$$
\quad (3.46)
$$

Here

$$
X_{kl,p} = (\tilde{k}_H + \tilde{l}_H + \tilde{p}_H)^2 \frac{(p_3^2 - k_3 l_3)^2}{16\, k\, l\, p\, k_H\, l_H\, p_H} \left(\frac{\tilde{p}_H^2 - \tilde{k}_H \tilde{l}_H}{p_3^2 - k_3 l_3} p_3 - \frac{\tilde{l}_H^2}{l_3} - \frac{\tilde{k}_H^2}{k_3} \right)^2, \quad (3.47)
$$

$$
Y_{kl,p}^\pm = (\tilde{k}_H - \tilde{l}_H + \tilde{p}_H)^2 \frac{(k_3^2 + l_3 p_3)^2}{16\, k\, l\, p\, k_H\, l_H\, p_H} \left(\frac{\tilde{k}_H^2 \pm \tilde{l}_H \tilde{p}_H}{k_3^2 + l_3 p_3} k_3 - \frac{\tilde{p}_H^2}{p_3} \pm \frac{\tilde{l}_H^2}{l_3} \right)^2 \quad (3.48)
$$

and $\tilde{k}_H = sign[k_3]k_H$.

We will look for horizontally isotropic and vertically averaged stationary solutions of (3.46) by assuming that wavenumber densities N_p depend neither on orientation of the horizontal wavenumber in the horizontal plane nor on the sign of the vertical wavenumber. In this case we may perform explicitly all angular integrations in the r.h.s. of the kinetic equation and, by assuming a scaling form for solutions

$$
N_p \sim p_H^\alpha\, |p_3|^\beta \quad (3.49)
$$

we get:

$$< \dot{N}(p_H, p_3) > \sim \int dk_H \, dl_H \, dk_3 \, dl_3 \; \delta(k_3 + l_3 - p_3) \frac{k_H l_H}{\triangle} \{$$

$$X_{kl,p} \left[k_H^\alpha \, l_H^\alpha \, k_3^\beta \, l_3^\beta - k_H^\alpha \, p_H^\alpha \, k_3^\beta \, p_3^\beta - l_H^\alpha \, p_H^\alpha \, l_3^\beta \, p_3^\beta \right]$$

$$\times \left[1 - (\frac{p_H}{k_H})^{2\alpha+6}(\frac{p_3}{k_3})^{2\beta+2} - (\frac{p_H}{l_H})^{2\alpha+6}(\frac{p_3}{k_3})^{2\beta+2} \right] \delta\left(\frac{k_H}{k3} + \frac{l_H}{l_3} - \frac{p_H}{p_3} \right) \} +$$

$$2 \int dk_H \, dl_H \, dk_3 \, dl_3 \; \delta(k_3 + l_3 - p_3) \frac{k_H l_H}{\triangle} \{$$

$$Y_{kl,p}^+ \left[k_H^\alpha \, l_H^\alpha \, k_3^\beta \, l_3^\beta + k_H^\alpha \, p_H^\alpha \, k_3^\beta \, p_3^\beta - l_H^\alpha \, p_H^\alpha \, l_3^\beta \, p_3^\beta \right]$$

$$\times \left[1 - (\frac{p_H}{k_H})^{2\alpha+6}(\frac{p_3}{k_3})^{2\beta+2} + (\frac{p_H}{l_H})^{2\alpha+6}(\frac{p_3}{l_3})^{2\beta+2} \right] \delta\left(\frac{l_H}{l_3} + \frac{p_H}{p_3} - \frac{k_H}{k_3} \right) \} \quad (3.50)$$

with $\triangle = \frac{1}{4}[2(p_H^2 k_H^2 + p_H^2 l_H^2 + l_H^2 k_H^2) - p_H^4 - k_H^4 - l_H^4]^{\frac{1}{2}}$

A direct substitution shows that there exist two sets of scaling exponents providing solutions. One annihilating the first bracket is a Rayleigh- Jeans type solution $\alpha = -1$, $\beta = 1$ leading to the equipartition of energy:

$$N_k \sim \frac{k_3}{k_H}, \qquad \epsilon_k = const \qquad (3.51)$$

One annihilating the second bracket is a Kolmogorov-type solution $\alpha = -\frac{7}{2}$, $\beta = -\frac{1}{2}$ giving

$$N_k \sim k_H^{-\frac{7}{2}} k_3^{-\frac{1}{2}}, \qquad \epsilon_k \sim k_H^{-\frac{5}{2}} k_3^{-\frac{3}{2}}. \qquad (3.52)$$

A simple check of dimensions shows that it corresponds to a constant energy flux through the wave spectrum

3.3 Historical remarks and bibliography

The presentation in this section follows the paper Caillol and Zeitlin (2000) where the kinetic equations in Eulerian variables were obtained and power-law spectra for internal gravity waves were first derived. Power-law spectra for *unidirectional* internal gravity waves were derived in Daubner and Zeitlin (1996) . The weak turbulence ideas were first applied to the internal gravity waves in the stratified fluid in Pelinovsky and Raevsky (1977) by using the Clebsch variables. There is a huge literature on observed atmospheric and oceanic spectra, cf. e.g. Bacmeister *et al* (1996) for the atmosphere. The resonant interactions and kinetics of the internal waves in the ocean were intensively discussed in the context of theoretical explanation of the celebrated empirical Garrett - Munk spectrum, cf. e.g. review Muller *et al* (1976) and references therein.

There are, however, some pecularities in applying Hamiltonian formalism to non-canonical Hamiltonian system, such as (3.1) which were not adressed in the earlier studies. Lagrangian variables used, e.g. in Olbers (1976) and other works, are subject to incompressibility constraint and are not canonical, which makes their use rather tricky.

The search of exact power-law solutions of the kinetic equations was never undertaken either in the early works.

Bibliography

J. T. Bacmeister *et al*, *J. Geoph. Res. D*, **5** 101: 9441–9441, 1996.

A.M. Balk and S.V. Nazarenko, Sov. Phys. JETP, 97: 1827–1845, 1990.

D.J. Benney and P.G. Saffman, Proc. R. Soc. A, 289: 301–320, 1966.

D.J. Benney and A.C. Newell, Stud. Appl. Math., 48: 29–35, 1969.

P. Caillol and V. Zeitlin, Dyn. Atmos. Oceans, 32: 81–112, 2000, Erratum: 33: 325–326, 2000.

S. Daubner and V. Zeitlin, Phys. Letters A, 214: 33–39, 1996.

F.E. Falkovich and Medvedev, S.B. Europhys. Lett., 19: 279–284, 1992.

A.A. Galeev and V.I. Karpman, Sov. Phys. JETP, 44: 592, 1963.

K. Hasselmann, Proc. R. Soc. A, 299: 77-100, 1967.

B.B. Kadomtsev and V.M. Kontorovich, *The theory of turbulence in hydrodynamics and plasma*, Sov. Phys. Radiophysics, 18: 511–540, 1974.

A.V. Kats and V.M. Kontorivich, V.M., Sov. Phys. JETP, 37: 80–85; 1974 (Sov. Phys. JETP, 38: 102–107, 1973).

K. Kenyon, *Notes on the 1964 Summer Study Programme on GFD at WHOI*, 11; 69, 1964.

E.A. Kuznetsov, Sov. Phys. JETP, 35; 310–314, 1972.

J. Le Sommer, G.M. Reznik, and V. Zeitlin, J. Fluid. Mech., 515: 35–170, 2004.

M. Longuet-Higgins and A.E. Gill, Proc. R. Soc. A, 299: 120-140, 1967.

S.B. Medvedev and V. Zeitlin, Phys. Letters A, *submitted*, 2004.

A.S. Monin and L.I. Piterbarg, Sov. Phys. Doklady, 32: 622–624, 1987.

P. Müller *et al*, Rev. Geophys., 24: 493 – 536, 1986.

D. Olbers, J. Fluid Mech., 74: 375–379, 1976.

E.N. Pelinovsky and M.A. Raevsky, Atm. Ocean Phys. - Izvestija, 13: 187–193, 1977.

G.M. Reznik and T.E. Soomere, Sov. Phys. Oceanology, 23:923–927, 1983.

G.M. Reznik, Sov. Phys. Oceanology, 25: 869–873, 1984.

G.M. Reznik,, V. Zetlin and M. Ben Jelloul, J. Fluid Mech., 445: 93–120, 2001.

R.Z. Sagdeev and A.A. Galeev, *Nonlinear plasma theory*, Benjamin, NY, 1969.

S.V. Volotsky, A.V. Kats, and V.M. Kontorovich, Proceedings of the Ukranian Akademy of Science (in Russian), 11: 66–69, 1980.

G.M. Zaslavsky and R.Z. Sagdeev, Sov. Phys. JETP, 52: 1081, 1967.

V.E. Zakharov, *Kolmogorov spectra in weak turbulence problems*, in "Handbook of Plasma Physics", Rozenbluth, M.N., Sagdeev, R.Z., Eds, 2: 3 – 36, 1984.

V.E. Zakharov and N.N. Filonenko, Sov. Phys. Doklady, 170: 1292, 1966.

V.E. Zakharov and L.I. Piterbarg, Sov. Phys. Doklady, 32: 560, 1987.

V.E. Zakharov, V.S. L'vov and G. Falkovich, *Kolmogorov spectra of turbulence I*, Springer, Berlin, 1992.

V. Zeitlin, Phys. Lett. A, 164: 177–183, 1992.

Nonlinear Amplitude Equations and Soliton Excitations in Bose-Einstein Condensates

Guoxiang Huang

Department of Physics, East China University, Shanghai 200062, China

Abstract We consider the soliton excitations in Bose-Einstein condensates (BECs) with a repulsive interparticle interaction. We show that long wavelength nonlinear excitations can be described by the Korteweg-de Vries equation in a cigar-shaped BEC and by the Kadomtsev-Petviashvili equation in a disk-shaped BEC. The nonlinear excitations with a short wavelength in a disk-shaped BEC obey the Davey-Stewartson equations. We also show that it is possible to realize a second harmonic generation of the nonlinear excitations in a two-component BEC.

1 Introduction

Microscopic particles can be classified as bosons, which have integer spins and obey Bose-Einstein statistics, and fermions, which have half-odd-integer spins and obey Fermi-Dirac statistics. At the temperature lowering than a critical value bosons have a tendency of attracting each other and occupy macroscopically a single quantum state, a phenomenon called Bose-Einstein condensation. Although the Bose-Einstein condensation was predicted by Einstein in 1925 [Einstein (1925)] and recognized as the theoretical basis of many macroscopic quantum phenomena, including superfluidity and superconductivity, the experimental demonstration of a Bose-Einstein condensate (BEC) for weakly interacting bosons was realized until 1995 [Anderson et al (1995), Davis et al (1995), Bradley et al (1995)], thanks to the development of the technology of laser cooling and trapping. Recently, molecular BECs have been realized for weakly interacting Fermi atomic dimers based on a Feshbach resonance technology [Jochim et al (2003), Greiner et al (2003), Zwierlein (2003)].

Because the Bose-Einstein condensation occurs at low temperature, quantum cooperative effect of microscopic particles plays an important role. In a BEC all particles moves coherently and the condensate can be well described by a macroscopic wave function, called order parameter. Due to inter-particle interaction, the dynamics of matter waves in BECs display many interesting nonlinear features. The experimental realization of Bose-Einstein condensation in weakly interacting atomic and molecular gases [Anderson et al (1995), Davis et al (1995), Bradley et al (1995), Jochim et al (2003), Greiner et al (2003), Zwierlein (2003)] has opened a new avenue for the study of the nonlinear properties of matter waves. The most spectacular experimental progress achieved recently are the realization of atomic four-wave mixing [Deng et al (1999)], the observation of dark [Burger et al (1999), Denschlag et al (2000)] and bright [Strecker et al (2002), Khaykovich et al (2002)] solitons and vortices [Mattews et al (1999)], the discovery of matter-wave amplification and superradiance [Inouye et al (1999a), Inouye et al (1999b), Kozuma et al (1999)]. Motivated by these important experimental findings, a large amount of theoretical works in this field has emerged [Perez-Garcia et al (1998), Golstein et al (1999), Bush et al (2000), Trombettoni et al (2001), Burger et al (2002), Huang (2001), Huang et al (2001), Huang (2002), Huang et al (2003)], and new phenomena such as atom holography through BEC [Zobay et al (1999)], coherent matter-wave amplification and superradiance in degenerate Fermi gases [Moore et al (2001)], etc., have also been predicted. These researches on nonlinear matter waves based on BEC have enabled the extension of linear atom optics to a nonlinear regime, i. e., *nonlinear atom optics* [Rolston and Phillips (2002)], very much like the laser lead to the development of nonlinear optics in the 1960s.

In this lecture we are interested in the soliton excitations in the BECs with a repulsive interparticle interaction. Note that, mathematically, the definition of soliton is different from solitary wave, but here we do not make distinction between them. Note that the experimental observations of the solitons in BECs have been done by several research groups. The Ertmer group at Hannover [Burger et al (1999)] and Phillips group at NIST [Denschlag et al(2000)] observed the dark solitons in respectively a cigar-shaped BEC of ^{87}Rb and an almost sphere-shaped BEC of ^{23}Na. Both ^{87}Rb and ^{23}Na have a repulsive inter-particle interaction. Bright solitons and bright soliton trains in BECs have been observed by Salomon group at ENS-Paris [Strecker et al (2002)] and Hulet group at Rice University [Khaykovich et al (2002)] in a cigar-shaped BEC of ^{7}Li, which has an attractive inter-particle interaction. Our lecture will concentrate on the nonlinear amplitude equations and relevant soliton excitations in the BECs with a repulsive inter-particle interaction. In the next section we introduce the nonlinear wave equation controlling the evolution of the macroscopic wave function starting from the Heisenberg equation of motion of quantum field operators by using a Bogoliubov approximation. Section III and section IV give a detailed derivation of Korteweg-de Vries (KdV) and Kadomtsev-Pitvashvilli (KP) equations for the weakly nonlinear, long wavelength excitations created in BECs with a repulsive atom-atom interaction in respectively a cigar-shaped trap and a disk-shaped trap. In section V we consider short wavelength excitations in a disk-shaped trap and derive Davy-Stewartson (DS) equations. Section V investigates the second harmonic generation of the nonlinear excitations in a two-component BEC. The last sections contains a summary and a discussion of our results.

2 Gross-Pitaevskii equation

The grand canonical Hamiltonian of a weakly interacting bosons is given by [Dalfovo et al (1999), Leggett (2001), Pethick and Smith (2002)]

$$
\hat{H} = \int d^3\mathbf{r}\hat{\psi}^{\dagger}(\mathbf{r},t)\left[-\frac{\hbar^2}{2m}\nabla^2 + V_{ext}(\mathbf{r}) - \mu\right]\hat{\psi}(\mathbf{r},t)
$$

$$
+\frac{1}{2}\int d^3\mathbf{r}d^3\mathbf{r}'\hat{\psi}^{\dagger}(\mathbf{r},t)\hat{\psi}^{\dagger}(\mathbf{r}',t)V(\mathbf{r}-\mathbf{r}')\hat{\psi}(\mathbf{r}',t)\hat{\psi}(\mathbf{r},t), \qquad (2.1)
$$

where $\hat{\psi}(\mathbf{r}, t)$ and $\hat{\psi}^\dagger(\mathbf{r}, t)$ are the boson field operators that annihilate and create a boson at location $\mathbf{r}$ and time t, respectively. m is the mass of particle, μ is the chemical potential and $V(\mathbf{r}-\mathbf{r}')$ is two-body interparticle potential. $V_{ext}(\mathbf{r})$ is an external trapping potential with the following form in most of BEC experiments [Dalfovo et al (1999), Leggett (2001)]

$$V_{ext}(\mathbf{r}) = \frac{m}{2}[\omega_x^2 x^2 + \omega_y^2 y^2 + \omega_z^2 z^2)], \tag{2.2}$$

where ω_j $(j = x, y, z)$ is the frequency of the trap in the jth direction. According to quantum mechanics, the field operator satisfies the Heisenberg equation of motion

$$i\hbar\frac{\partial}{\partial t}\hat{\psi}(\mathbf{r}, t) = \left[\hat{\psi}, \hat{H}\right]$$
$$= \left[-\frac{\hbar^2}{2m}\nabla^2 + V_{ext}(\mathbf{r}) - \mu + \int d^3\mathbf{r}'\hat{\psi}^\dagger(\mathbf{r}', t)V(\mathbf{r}' - \mathbf{r})\hat{\psi}(\mathbf{r}', t)\right]\hat{\psi}(\mathbf{r}, t), \tag{2.3}$$

with the canonical commutation relations

$$\left[\hat{\psi}(\mathbf{r}, t), \hat{\psi}^\dagger(\mathbf{r}', t)\right] = \delta(\mathbf{r} - \mathbf{r}'), \tag{2.4}$$

$$\left[\hat{\psi}(\mathbf{r}, t), \hat{\psi}(\mathbf{r}', t)\right] = \left[\hat{\psi}^\dagger(\mathbf{r}, t), \hat{\psi}^\dagger(\mathbf{r}', t)\right] = 0. \tag{2.5}$$

In a dilute cold gas at low energy only binary collisions are important and these collisions can be characterized by a single parameter, independent of the details of the two-body potential. This allows us to use the contact potential approximation $V(\mathbf{r}-\mathbf{r}') = g\delta(\mathbf{r}' - \mathbf{r})$, where $g = 4\pi\hbar^2 a_s/m$. The parameter a_s is called s-wave scattering length, which may be positive and negative, corresponding repulsive and attractive interparticle interaction, respectively [Dalfovo et al (1999), Leggett (2001), Pethick and Smith (2002)]. At low temperature most of particles are in the condensate one can use the Bogoliubov approximation

$$\hat{\Psi}(\mathbf{r}, t) = \psi(\mathbf{r}, t) + \hat{\psi}'(\mathbf{r}, t), \tag{2.6}$$

where $\psi(\mathbf{r}, t)$ and $\hat{\psi}'(\mathbf{r}, t)$ describe respectively condensed and thermal components of the system. For a dilute gas at zero temperature one can neglect safely the thermal component [Dalfovo et al (1999), Leggett (2001), Pethick and Smith (2002)], and hence Eq. (2.3) under the contact approximation is reduced to

$$i\hbar\frac{\partial\psi}{\partial t} = \left[-\frac{\hbar^2}{2m}\nabla^2 + V_{ext}(\mathbf{r}) - \mu + g|\psi|^2\right]\psi, \tag{2.7}$$

called Gross-Pitaevskii (GP) equation [Dalfovo et al (1999), Leggett (2001), Pethick and Smith (2002)]. The function ψ is called condensed state wavefunction satisfying the condition $\int d^3\mathbf{r}|\psi|^2 = N$, where N is the particle number in the condensate.

3 KdV equation for a cigar-shaped BEC

3.1 Hydrodynamic form of the GP equation

We first consider the nonlinear excitations created in a cigar-shaped BEC with a repulsive interparticle interaction. We shall show that the shallow dark solitons observed by Ertmer group [Burger (1999)] can be well described by the KdV equation [Huang (2001), Huang (2001a)]. Expressing the order parameter in terms of its modulus and phase, $\psi = \sqrt{n} \exp(i\phi)$, we obtain a set of coupled equations for n and ϕ. By introducing $(x, y, z) = a_z(x', y', z')$, $t = \omega_z^{-1} t'$, $n = n_0 n'$ with $a_z = \sqrt{\hbar/(m\omega_z)}$ and $n_0 = N/a_z^3$, Eq. (2.7) takes the following dimensionless hydrodynamic form (after dropping the primes):

$$\frac{\partial n}{\partial t} + \nabla \cdot (n \nabla \phi) = 0, \tag{3.1}$$

$$\frac{\partial \phi}{\partial t} + V_x(x) + \frac{1}{2}\left[\left(\frac{\omega_y}{\omega_z}\right)^2 y^2 + z^2\right] + Qn + \frac{1}{2}\left[(\nabla\phi)^2 - \frac{1}{\sqrt{n}}\nabla^2\sqrt{n}\right] = 0, \tag{3.2}$$

with $V_x(x) = (\omega_x/\omega_z)^2 x^2/2$, $Q = 4\pi N a/a_z$ and $\int d^3\mathbf{r} n = 1$. The last term on the left hand side of Eq. (3.2) (i.e. $-\frac{1}{2\sqrt{n}}\nabla^2\sqrt{n}$) is called the quantum pressure, which provides the the dispersion necessary to form a dark soliton in the BEC as we shall see below.

We are interested in the excitation created in the condensate with a cigar-shaped trap. The cigar-shaped trap here means that the trapping frequency in one direction is much less than the other two directions, e. g. the conditions $\omega_x \ll \omega_y, \omega_z$ ($\omega_y \approx \omega_z$), $a_z \ll l_0$ and $\hbar\omega_x \ll n_0 g \ll \hbar\omega_z$ can be fulfilled, where $l_0 = (4\pi n_0 a_s)^{-1/2}$ is called healing length. For simplicity we here assume $\omega_y = \omega_z$ and take $\omega_x = 0$. In this situation two consequences follow: (i) The energy-level spacing in the y and z directions exceeds largely the interaction energy between particles, and hence the condensate is quasi-one-dimensional (1D). Thus at sufficiently low temperature the motion of particles in the y and z directions is essentially frozen and is governed by the ground-state wave function (zero-point oscillation) of the corresponding harmonic oscillator [Pethick and Smith (2002)]. (ii) Due to the strong confinement in the y and z directions, the excitations can propagate *only* in the axial (i. e. in the x) direction, similar to an electromagnetic wave propagating along an optical fiber [Agrawal (2001)]. Consequently, the superfluid velocity $\mathbf{v}$ ($= \nabla\phi$) has only x component, and hence $\phi = \phi(x, t)$.

Based on the above observations, we set $\sqrt{n} = P(x, t)G_0(y, z)$, $\phi = -\mu t + \varphi(x, t)$, or equivalently $\psi(x, y, z, t) = G_0(y, z)P(x, t)\exp[-i\mu t + i\varphi(x, t)]$, where $G_0(y, z) = \exp[-(y^2 + z^2)/2]$ is the ground-state wave function of the 2D harmonic oscillator with the potential $(y^2 + z^2)/2$, μ is the chemical potential of the condensate and φ is a phase function contributed from the excitation, which is assumed to be a function of x because as mentioned above the excitation can only propagate in the x direction.

With these considerations Eqs. (3.1) and (3.2) are transformed into (1+1)-D form

$$\frac{\partial P}{\partial t} + \frac{\partial P}{\partial x}\frac{\partial \varphi}{\partial x} + \frac{1}{2}P\frac{\partial^2 \varphi}{\partial x^2} = 0, \tag{3.3}$$

$$-\frac{1}{2}\frac{\partial^2 P}{\partial x^2} + (-\mu + 1)P + \left[\frac{\partial \varphi}{\partial t} + \frac{1}{2}\left(\frac{\partial \varphi}{\partial x}\right)^2\right]P + \frac{1}{2}QP^3 = 0. \tag{3.4}$$

To obtain Eq. (3.4) we have used Eq. (3.2) with $G = G_0$ and multiplied Eq. (3) by G_0^* and then integrated once with respect to y and z. This projection procedure is equivalent to assume that the excitation is a quasi-1D one. [Huang (2001), Huang (2001a), Huang (2002)].

3.2   Linear dispersion relation of the excitations

Because there is no trapping potential in the axial direction, the ground state wave function in this direction is a constant, say u_0. To consider linear excitations from the ground state we take $P = u_0 + a(x, t)$, where $a(x, t)$ represents the excitations. Assuming that

$$(a, \varphi) = (a_0, \varphi_0) \exp[i(kx - \omega t)] + c.c., \tag{3.5}$$

with a_0 and φ_0 being constants, we obtain the linear dispersion relation of Eqs. (3.3) and (3.4)

$$\omega(k) = \pm \frac{1}{2} k \left(2Qu_0^2 + k^2 \right)^{1/2}, \tag{3.6}$$

where the positive (respectively, negative) sign corresponds to the wave propagating to the right (respectively, left). We stress that the k^2-term in the bracket of Eq. (3.6) comes from the quantum pressure, given by the term $-(1/2)\partial^2 P/\partial x^2$ in Eq. (3.4). Equation (3.6) is the Bogoliubov-type linear excitation spectrum of our system for sound propagation. We see that to get the Bogoliubov excitation spectrum, the quantum pressure of the system plays a significant role. From (3.6) we obtain the sound speed of the system $c = (d\omega/dk)|_{k=0} = \pm\sqrt{Q/2}u_0$. For an homogeneous system (i.e. $V_{ext}(\mathbf{r}) = 0$) the corresponding sound speed is $c_0 = \pm\sqrt{Q}u_0$ in our notation. Thus we have $c/c_0 = 1/\sqrt{2}$. The factor $1/\sqrt{2}$ is due to the transverse confinement of the system. This result is consistent with the experiment by Andrews et al [Andrews (1998)].

3.3   KdV equation for weakly nonlinear excitations

Now we consider the weakly nonlinear, long wavelength excitations from the condensate. Using the asymptotic expansion

$$P = u_0 + \epsilon^2(a^{(0)} + \epsilon^2 a^{(1)} + \cdots), \tag{3.7}$$

$$\varphi = \epsilon(\varphi^{(0)} + \epsilon^2 \varphi^{(1)} + \cdots), \tag{3.8}$$

and assuming that $a^{(j)}$ and $\varphi^{(j)} (j = 0, 1, \cdots)$ are functions of the multiple-scale variables $\xi = \epsilon(x - ct)$ and $\tau = \epsilon^3 t$, where ϵ is a smallness parameter characterizing the relative amplitude of the excitation, and then substituting them to Eqs. (3.3) and (3.4), we obtain

$$c \frac{\partial a^{(j)}}{\partial \xi} - \frac{1}{2} u_0 \frac{\partial^2 \varphi^{(j)}}{\partial \xi^2} = \alpha^{(j)}, \tag{3.9}$$

$$Q u_0^2 a^{(j)} - c u_0 \frac{\partial \varphi^{(j)}}{\partial \xi} = \beta^{(j)}. \tag{3.10}$$

The explicit expressions of $\alpha^{(j)}$ and $\beta^{(j)}$ ($j = 0, 1, \cdots$) are omitted here.

In the leading order ($j = 0$) we obtain $\varphi^{(0)} = (2c/u_0) \int d\xi a^{(0)}$ with $a^{(0)}$ a function yet to be determined. The solvability condition demands $c = \delta_1 \sqrt{Q/2}u_0$ with $\delta_1 = \pm 1$. At the next order ($j = 1$), the solvability condition results in the closed equation for $a^{(0)}$:

$$\frac{\partial a^{(0)}}{\partial \tau} + \frac{3c}{u_0} a^{(0)} \frac{\partial a^{(0)}}{\partial \xi} - \frac{1}{8c} \frac{\partial^3 a^{(0)}}{\partial \xi^3} = 0. \tag{3.11}$$

Equation (12) is the KdV equation for the wave travelling to the right (left) for the case $\delta_1 = +1$ $(\delta_1 = -1)$. Let $w = \epsilon^2 a^{(0)}$ and use the definition of ξ and τ we obtain

$$\frac{\partial w}{\partial t} + \frac{3c}{u_0} w \frac{\partial w}{\partial X} - \frac{1}{8c} \frac{\partial^3 w}{\partial X^3} = 0, \tag{3.12}$$

with $X = x - ct$. By the scaling transformation $w = u_0 \tilde{w}/(4c^2)$ and $t = -8c\tilde{t}$ Eq. (3.12) is reduced to the standard form $\partial \tilde{w}/\partial \tilde{t} - 6\tilde{w}\partial \tilde{w}/\partial X + \partial^3 \tilde{w}/\partial X^3 = 0$, studied intensively in soliton theory and applied widely in many fields [Grimshaw (2004)].

3.4 Dark solitons and their interaction

The single-soliton solution of Eq.(3.12) is given by

$$w = -W_0 \operatorname{sech}^2 \left[\sqrt{\frac{2c^2 W_0}{u_0}} \left(X + c\frac{W_0}{u_0}t - x_0 \right) \right], \tag{3.13}$$

where W_0 is a *positive* constant, x_0 is a constant denoting the initial position of the soliton on the pedestal background. Exact to the first order, the condensed-state wave function takes the form

$$\psi = \quad u_0 \left(1 - \widetilde{W_0} \operatorname{sech}^2 \left[\sqrt{2c^2 \widetilde{W_0}} \left(x - c(1 - \widetilde{W_0})t - x_0 \right) \right] \right)$$

$$\times \exp \left[-\frac{y^2 + z^2}{2} \right] \exp[i(-\mu t + \varphi)], \tag{3.14}$$

with $\widetilde{W_0} = W_0/u_0$, $\mu = 1 + Qu_0^2/2$. The constant $\widetilde{W_0}$ can be taken as the grayness of the soliton. The phase function reads

$$\varphi = -\sqrt{2\widetilde{W_0}} \tanh \left[\sqrt{2c^2 \widetilde{W_0}} \left(x - c(1 - \widetilde{W_0})t - x_0 \right) \right]. \tag{3.15}$$

From (3.14) we see that the excitation is a *dark* soliton (i.e. density minimum of the condensate) and its propagating velocity is $v_s = c(1 - \widetilde{W_0})$, lower than the sound speed c of the system. The formation of the *dark* soliton is due to the balance between the nonlinearity and the dispersion. Consequently, the quantum pressure plays a significant role in the formation of the dark soliton in the BEC. Since the velocity of the dark soliton may have two different signs, i.e. it may propagate in either the positive or the negative x-direction and thus a head on collision of two dark soliton in the BEC is possible.

In above calculation, the slowly-varying trapping in the axial direction is neglected for simplicity. In fact one can develop a systematic analytical approach to consider the effect of the axial trapping potential. Huang et al (2002) shows that, for a condensate strongly confined in two transverse directions, the ground state of the system involves the high-order eigen-modes of the transverse confining potential in the transverse directions and effective high-order Thomas-Fermi wave functions in the axial direction. The linear excitations of the system have a Bogoliubov-type spectrum with the excitation frequency varying slowly along the axial direction. In weak nonlinear approximation the amplitude of a nonlinear excitation is governed by a variable coefficient KdV equation with additional terms contributed from the transverse structure and the inhomogeneity in the axial direction of the condensate, which results in varying amplitude, width and velocity for dark solitoins. Because of the inhomogeneity the dark solitons undergo deformation and emit radiations when travelling along the axial direction. A dark soliton will disintegrate into several ones plus a residual wavetrain when passing over a step-like potential [Huang et al (2002)].

An interesting dynamic property for solitons is their interaction. In a 1D- (or quasi-1D) system, there are two distinct soliton interactions. One is the *overtaking* collision and the other one is the *head-on* collision. [Zabusky and Kruskal (1965), Su and Mirie (1980), Linde et al (1993)] The overtaking collision of the dark solitons in the BEC can be studied with the KdV Eq. (3.12). Its multi-soliton solutions (they travel in the same direction) can be obtained from the inverse scattering transform [Drazin and Johnson (1989)] . However, for the head-on collision we must search for the evolution of waves travelling to both sides and hence we need to employ a suitable asymptotic expansion to solve the original equations of motion (3.1) and (3.2), or their simplified form (3.3) and (3.4) [Su and Mirie (1980)]. A detailed analytical and numerical calculation can be found in Huang et al (2001a). Here we just present some numerical results for the head-on collisions of two dark solitons in the BEC.

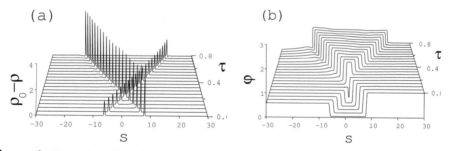

Figure 1. Space-time plots of two colliding solitons. (a) relative amplitude, (b) phase.

Shown in Fig.1 is the result of head-on collision of two solitons. Initially two solitons are created in the condensate. They propagate in opposite directions, approach each other, collide and, asymptotically, separate away. Due to the axial inhomogeneity the solitons disply some distortions during collision. In the figure $s = (\omega_x/\omega_z)^{1/2}$ and $\tau = [\omega_x/(2\omega_z)]t$. ρ is a new variable proportional to the amplitude of the order parameter ψ and ρ_0 is its stationary value.

Two different head-on collisions are found(Fig. 2): (i) *"Gray collision"* (Fig. 2a). If the initial soliton amplitudes or depression depths are small enough, during collision a single composite structure forms and further increases its amplitude but never touches zero, i.e. waves remain always gray. Figure 2c shows the paths of the corresponding solitons. Clearly, around $\tau = 0.2$ the solitons form a single composite solution, which survives during some time interval (vertical bar linking two parabolic-like paths). Then,

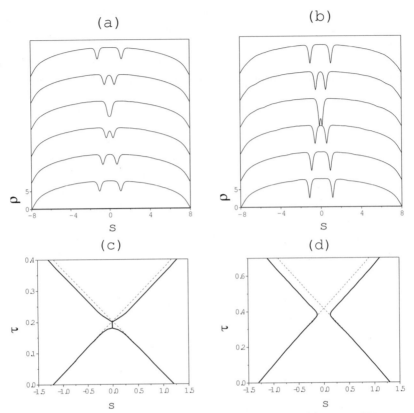

Figure 2. Two different types of soliton collisions. (a), (c) Gray collision; (b), (d) Black collision. (a, b) Sequential snapshots of amplitude distribution, (c, d) soliton paths (only a piece, $s \in [-1.5, 1.5]$, of the whole interval is plotted). Dotted lines mark soliton path as if each of them would propagate alone.

the solitons separate away. (ii) *"Black collision"* (Fig. 2b). If the initial soliton amplitudes are large enough, then, while the solitons approach each other, their amplitudes grow and at some instant of time, just before collision, the soliton amplitudes vanish as their corresponding depressions touch zero (solitons become plain black) still remaining on some distance from each other (fourth snapshot on Fig. 2b). Thus in this case the solitons never form one hump solution (Fig. 2d). In both cases the soliton collisions show positive position shifts (phase shifts), see Fig.2c and Fig.2d.

4 KP equation for a disk-shaped BEC

Although elongated cigar-shaped traps have been widely used in BEC experiments, a flat disk-shaped trap, in which the confinement in one direction is much larger than other two directions, has also been employed [Jin et al (1996)]. In fact, the JILA trap, which was used by Anderson *et al* [Anderson et al (1995)] for the first experimental observation of the Bose-Einstein condensation of weakly interacting alkali atomic gases, is just of this type. Later on a disk-shaped trap was also used by Jin *et al* [Jin et al (1996)] to investigate the phonon-like linear excitations in BECs. If the thickness of the disk is small enough, the condensate becomes quasi-2D. One expects that at sufficiently low temperature the motion of atoms in the direction perpendicular to the disk is frozen and governed by the ground-state wave function in that direction. Such quasi-2D BEC has recently been realized experimentally by Görlitz *et al* (2001). In this section we investigate weakly nonlinear, long wavelength excitations in a disk-shaped trap. We obtain a KP equation, i. e. the 2D generalization of the KdV equation.

We consider an anisotropic harmonic trap of the form $V_{ext}(\mathbf{r}) = \frac{m}{2}[\omega_\perp^2(x^2+y^2)+\omega_z^2 z^2]$ with $\omega_\perp << \omega_z$. The dimensionless hydrodynamic equations of motion (the primes are omitted) take the form

$$\frac{\partial n}{\partial t} + \nabla \cdot (n\nabla\phi) = 0, \tag{4.1}$$

$$\frac{\partial \phi}{\partial t} + \frac{1}{2}z^2 + V_\parallel(x,y) + Qn + \frac{1}{2}\left[(\nabla\phi)^2 - \frac{1}{\sqrt{n}}\nabla^2\sqrt{n}\right] = 0, \tag{4.2}$$

with $Q = 4\pi N a_s/a_z$ and $\int d\mathbf{r}n = 1$. $V_\parallel(x,y) = (\omega_\perp/\omega_z)^2(x^2+y^2)/2$ is the dimensionless trapping potential in the x and y directions.

A disk-shaped trap means $\omega_\perp << \omega_z$. For simplicity we assume $\omega_\perp = 0$. Obviously in this case an excitation in the BEC can propagate in x and y directions and in z direction it is a standing wave. As in Sec. 3.1 we can take $\sqrt{n} = P(x,y,t)G_0(z)$ and $\phi = -\mu t + \varphi(x,y,t)$, i. e. equivalently $\psi(x,y,z,t) = G_0(z)P(x,y,t)\exp[-i\mu t + i\varphi(x,y,t)]$, where $G_0(z) = \exp(-z^2/2)$ is the ground-state wave function of the 1D harmonic oscillator with the potential $z^2/2$. μ is the chemical potential of the condensate and φ is a phase function contributed by excitation, assumed to be a function of x and y because the excitation can only propagate in the x and y directions.

Thus, Eqs. (4.1) and (4.2) can be reduced to

$$\frac{\partial P}{\partial t} + \frac{\partial P}{\partial x}\frac{\partial \varphi}{\partial x} + \frac{\partial P}{\partial y}\frac{\partial \varphi}{\partial y} + \frac{P}{2}\left(\frac{\partial^2\varphi}{\partial x^2} + \frac{\partial^2\varphi}{\partial y^2}\right) = 0, \tag{4.3}$$

$$-\frac{1}{2}\left(\frac{\partial^2 P}{\partial x^2} + \frac{\partial^2 P}{\partial y^2}\right) - (\mu - \frac{1}{2})P \tag{4.4}$$

$$+\left[\frac{\partial\varphi}{\partial t} + \frac{1}{2}\left(\frac{\partial\varphi}{\partial x}\right)^2 + \frac{1}{2}\left(\frac{\partial\varphi}{\partial y}\right)^2\right]P + Q'P^3 = 0,$$

where $Q'=I_0 Q$ is an effective interaction constant with $I_0 = \int_{-\infty}^\infty dz G_0^4(z)/\int_{-\infty}^\infty dz G_0^2(z) = 1/\sqrt{2}$. To arrive at Eq. (4.4) we have multiplied Eq. (4.2) by G_0^* and then integrated once with respect to z to eliminate the dependence on z. In principle, one can take into account the contribution of the higher-order eigen-modes of the harmonic oscillator in the z-direction, as done by Huang et al (2002) for a cigar-shaped trap. However, as here we have assumed $n_0 g \ll \hbar \omega_z$, the contribution from these higher-order eigen-modes is small and can be safely neglected.

The linear dispersion relation of the system reads

$$\omega(k_1, k_2) = \pm \frac{1}{2} k \left(4Q'u_0^2 + k^2\right)^{1/2}, \qquad (4.5)$$

where $k = (k_1^2 + k_2^2)^{1/2}$ is the wave number and ω is the frequency of the excitation. Equation (4.5) corresponds to a Bogoliubov-type linear excitation spectrum in 2D. The sound speed of the system is given by

$$c = \lim_{k \to 0} \left[\left(\frac{\partial \omega}{\partial k_1}\right)^2 + \left(\frac{\partial \omega}{\partial k_2}\right)^2\right]^{1/2} = \sqrt{Q'} u_0. \qquad (4.6)$$

Note that for a homogeneous system (i.e. $V_{ext}(\mathbf{r}) = 0$) the corresponding sound speed in our notation is $c_0 = \sqrt{Q} u_0$. Thus we have $c/c_0 = \sqrt{Q'/Q} = 1/2^{1/4}$.

To get a nonlinear amplitude equation for weakly nonlinear, long wavelength excitation we make the asymptotic expansion $P = u_0 + \epsilon^2(a^{(0)} + \epsilon^2 a^{(1)} + \cdots)$ and $\varphi = \epsilon(\varphi^{(0)} + \epsilon^2 \varphi^{(1)} + \cdots)$, where $a^{(j)}$ and $\varphi^{(j)}$ $(j = 0, 1, \ldots)$ are functions of the multiple-scale (slow) variables $\xi = \epsilon(c^{-1}x - t)$, $\eta = \epsilon^2 y$ and $\tau = \epsilon^3 t$. Substituting the expansion into Eqs. (4.3) and (4.4) we obtain

$$\frac{\partial a^{(j)}}{\partial \xi} - \frac{1}{2c^2} u_0 \frac{\partial^2 \varphi^{(j)}}{\partial \xi^2} = \alpha^{(j)}, \qquad (4.7)$$

$$2Q'u_0^2 a^{(j)} - u_0 \frac{\partial \varphi^{(j)}}{\partial \xi} = \beta^{(j)}, \qquad (4.8)$$

for $j = 0, 1, \ldots$ The explicit expressions of $\alpha^{(j)}$ and $\beta^{(j)}$ are omitted here.

In the leading order $(j = 0)$ we obtain $\varphi^{(0)} = (2c^2/u_0) \int d\xi\, a^{(0)}$ with $a^{(0)}$ a function yet to be determined. The solvability condition in this order requires $c = \sqrt{Q'} u_0$, which is just the sound speed of the system. At the next order $(j = 1)$, the solvability condition results in a closed equation for $a^{(0)}$:

$$\frac{\partial}{\partial \xi}\left(\frac{\partial a^{(0)}}{\partial \tau} + \frac{3\sqrt{Q'}}{c} a^{(0)} \frac{\partial a^{(0)}}{\partial \xi} - \frac{1}{8c^2}\frac{\partial^3 a^{(0)}}{\partial \xi^3}\right) + \frac{c^2}{2}\frac{\partial^2 a^{(0)}}{\partial \eta^2} = 0. \qquad (4.9)$$

Equation (4.9) is the KP equation [Grimshaw (2004)]. Note that (4.9) is a *positive-dispersion* KP equation (also called KP-I equation) since the dispersion term and the diffractive term (i.e. the second-order derivative-term with respect to η) have opposite signs.

The KP-I equation is a completely integrable system and can be solved by the inverse scattering transform [Ablowitz and Clarkson (1991)]. A plane soliton and lump (decaying in all directions) solutions can be obtained. By using these solutions as initial conditions and including the slowly-varying trapping potential $V_{\parallel}(x, y)$, a detailed numerical study has been done by Huang et al (2003) on the the stability of a plane dark soliton in a disk-shaped BEC. It is found that a low-depth plane *dark* soliton can propagate in the

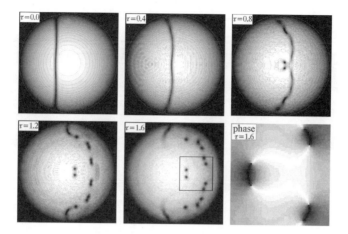

Figure 3. Time evolution of a high-depth dark plane soliton in a disk-shaped BEC. The soliton propagates from the left to the right with a developing snake instability eventually decaying into vortices. The last snapshot shows the phase in the square region marked on the corresponding amplitude snapshot for $\tau = 1.6$.

condensate with a changing profile but preserving its structure down to the boundary of the condensate. However, a high-depths plane dark soliton is unstable to long wavelength transverse disturbances. The instability appears as a longitudinal modulation of the soliton amplitude decaying into vortices (see Fig.3). The study also shows that a dark lump decaying algebraically in two spatial directions can propagate rather stable in the condensate but disappears near the boundary of the condensate where two vortices are nucleated. The vortices move in opposite directions along the boundary and when meeting merge creating a new lump (see Fig.4 and Fig.5).

When the slowly-varying trapping potential in the x and y directions is included, we can obtained a variable-coefficient KP equation with some additional terms. For detail, see Huang et al [Huang et al (2003)].

5 Davy-Stewartson equations

Note that the theoretical approaches developed in the last two sections are valid only for weak nonlinear excitations with a long wavelength and hence the dispersion of the excitations under consideration is weak. In fact, in addition to the long wavelength excitations, the BEC may support the nonlinear excitations with a shorter wavelength thus having a stronger dispersion. For the excitations with a strong dispersion one must employ a different multiple-scale expansion to get nonlinear envelope equations that describe the nonlinear dynamics of the system. In fluid dynamics and nonlinear optics, it is well known that such envelope equation is the nonlinear Schrödinger (NLS) equation in 1D and Davey-Stewartson (DS) equations in 2D [Craik (1985), Newell and Moloney (1992)]. Under some conditions the DS equations take the form of DSI equtions,

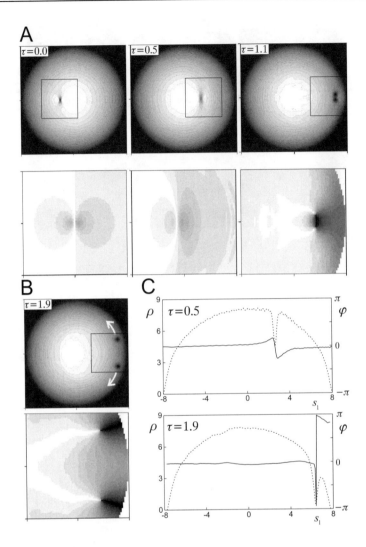

Figure 4. Propagation of a lump and the nucleation near the edge of two moving, clockwise and counterclockwise, vortices. A. Sequential snapshots of the condensate with a high-depth lump initially excited [in the first row brightness corresponds to amplitude value $\rho(s_1, s_2)$, in the second row brightness going from black to black via white reflects the phase $\varphi(s_1, s_2)$ in the interval $[-\pi, \pi]$ (enlarged square regions are shown), where $s_1 = (\omega_\perp/\omega_z)^{1/2}x$, $s_2 = (\omega_\perp/\omega_z)^{1/2}y$]. The lump propagates from the left to the right. Then, near the boundary, its depth approaches 100%, the lump becomes practically black and it breaks into two vortices. B. The vortices move along the boundary of the condensate in opposite directions (arrows show motion directions). C. Section of the condensate along the "horizontal" axis s_1 crossing the lump for $\tau = 0.5$ and the vortex for $\tau = 1.9$. Solid and dashed lines correspond to the phase and the amplitude, respectively.

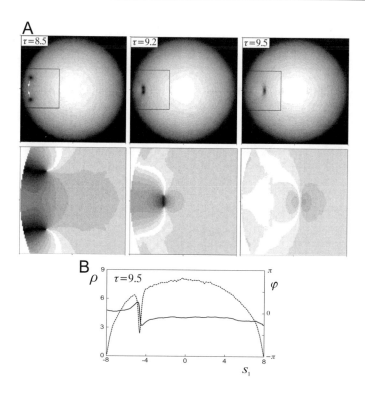

Figure 5. Formation of a new lump by the merging of two vortices [continuation of Fig. 4]. A. Sequential snapshots of the condensate before, during, and after the merging of the vortices. The lump formed in the process propagates from the left to the right very much like the initial one (Fig. 4). B. Section of the condensate along the "horizontal" axis s_1 crossing the lump for $\tau = 9.5$. (solid and dashed lines correspond to the phase and the amplitude, respectively).

which are completely integrable system and allow localized 2D soliton solutions decaying exponentially in all spatial directions.

In this section we study the dynamics of 2D weak nonlinear excitations with a short wavelength created in a BEC. We still assume that the interparticle interaction is repulsive and hence the system may display a Bogoliubov-type linear excitation spectrum. Although the GP equation (2.7) describing the whole condensate is of a cubic nonlinearity, the nonlinearity describing the interaction between the excitations is quadratic. We predict that a wavepacket superposed by short wavelength excitations will be coupled by a long wavelength mean flow which is generated by the wavepacket through an effect of the rectification of the short wavelength excitations.

As in the last section we consider a disk-shaped trap thus the 3D GP equation (2.7) is reduced to the Eqs. (4.3) and (4.4) with the linear dispersion relation (4.5). In order to investigate the weak nonlinear excitations with a short wavelength, following the line of Davey and Stewartson (1974) we make the asymptotic expansion

$$P = u_0 + \epsilon a^{(1)} + \epsilon^2 a^{(2)} + \epsilon^3 a^{(3)} + \cdots, \tag{5.1}$$

$$\varphi = \epsilon \varphi^{(1)} + \epsilon^2 \varphi^{(2)} + \epsilon^3 \varphi^{(3)} + \cdots, \tag{5.2}$$

and assume that $a^{(j)}$ and $\varphi^{(j)}$ $(j = 1, 2, 3, ...)$ are the functions of the fast variable $\theta = kx - \omega t$ and the slow variables $\xi = \epsilon(c_g^{-1}x - t)$, $\eta = \epsilon y$, $\tau = \epsilon^2 t$, where ϵ is a smallness parameter characterizing the relative amplitude of the excitation. c_g is a constant yet to be determined. Then substituting above expansion to Eqs.(4.3) and (4.4), we obtain

$$-\omega \frac{\partial a^{(j)}}{\partial \theta} + \frac{1}{2} u_0 k^2 \frac{\partial^2 \varphi^{(j)}}{\partial \theta^2} = \alpha^{(j)}, \tag{5.3}$$

$$\left[-\frac{1}{2} k^2 \frac{\partial^2}{\partial \theta^2} + 2Q' u_0^2 \right] a^{(j)} - \omega u_0 \frac{\partial \varphi^{(j)}}{\partial \theta} = \beta^{(j)}, \tag{5.4}$$

$j = 1, 2, 3, ...$ The explicit expressions of $\alpha^{(j)}$ and $\beta^{(j)}$ are omitted here.

At leading $(j = 1)$ order the solution reads

$$\varphi^{(1)} = A_0 + [A \exp(i\theta) + c.c.], \tag{5.5}$$

$$a^{(1)} = \frac{i}{2} \frac{u_0 k^2}{\omega} F \exp(i\theta) + c.c., \tag{5.6}$$

where A_0 is a mean flow necessarily to be introduced for cancelling a secular term appearing in high-order approximations. A is an envelope function of the carrier wave $\exp(i\theta)$. Both A_0 and A are yet to be determined functions of the slow variables ξ, η and τ. $\omega(k)$ is just the linear dispersion relation of the excitation, given in Eq. (4.5) with $k_1 = k$ and $k_2 = 0$. c.c. represents a corresponding complex conjugate term.

At the next order $(j = 2)$, a solvability condition requires that $c_g = d\omega/dk = [2Q'u_0^2 k + k^3]/(2\omega)$, i. e. the group velocity of the carrier wave. The singularity-free second order solution reads $\varphi^{(2)} = A_2 \exp(2i\theta) + c.c.$, and $a^{(2)} = B_{20} + [B_{21} \exp(i\theta) + B_{22} \exp(2i\theta) + c.c.]$, where A_2, B_{2j} $(j = 0, 1, 2)$ are functions of A_0, A, $\partial A_0/\partial \xi$ and $\partial A/\partial \xi$.

At the order $j = 3$ the solvability conditions give rise to the equations for A_0 and A:

$$\alpha_1 \frac{\partial^2 A_0}{\partial \xi^2} - \frac{\partial^2 A_0}{\partial \eta^2} = \alpha_2 \frac{\partial}{\partial \xi} \left(|A|^2 \right), \tag{5.7}$$

$$i \frac{\partial A}{\partial \tau} + \beta_1 \frac{\partial^2 A}{\partial \xi^2} + \beta_2 \frac{\partial^2 A}{\partial \eta^2} + \beta_3 |A|^2 A - \beta_4 A \frac{\partial A_0}{\partial \xi} = 0, \tag{5.8}$$

with the coefficients explicitly given by

$$\alpha_1 = \frac{1}{c^2} - \frac{1}{c_g^2},$$

$$\alpha_2 = \frac{k^2}{2c_p^3 c_g}\left(2c_p^2 + 3c_p c_g + c^2\right),$$

$$\beta_1 = \frac{1}{2wc_g^2}\left(c_p^2 - \frac{c^4}{c_p^2} + \frac{k^2}{4}\right),$$

$$\beta_2 = \frac{1}{2w}\left(c_p^2 + \frac{k^2}{4}\right),$$

$$\beta_3 = \frac{1}{2w}\left[c^2 k^2 + \left(\frac{15}{8}\frac{c^2}{c_p^2} - \frac{1}{4}\right)k^4 + k^2\left(1 + \frac{3}{2}\frac{c^2}{c_p^2}\right)\frac{c_p^2(2c^2 + k^2) + 3c^2 k^2/4}{c^2 + k^2}\right],$$

$$\beta_4 = \frac{1}{2wc_g}\left(c_g + 2c_p\right)k^2, \tag{5.9}$$

where c_p is phase speed defined by $w(k)/k$. We see that due to nonlinear effect a coupling occurs between the envelope of the short-wavelength excitations and the long-wavelength mean flow of the system. Such nonlinear coupling is one type of interactions between long waves and short waves.

Equations (5.7) and (5.8) are generalized Davey-Stewartson (DS) equations, which appear also in fluid physics, nonlinear optics, lattice dynamics, and plasma physics and have generated much interest in recent years [Craik (1985), Neweel and Moloney (1992), Cui et al (2003), Huang et al (2001b), Khismatulin and Akhatov (2001)].

We now investigate the 2D soliton solutions of the DS Eqs. (5.7) and (5.8). Using the transformation $\partial A_0/\partial\xi = -\epsilon^{-2}\beta_1 k^4/(\alpha_1\beta_4)s$, and $A = \epsilon^{-1}[4\beta_1/(\alpha_2\beta_4)]^{1/2} k^4 u$, Eqs. (5.7) and (5.8) can be rewritten as the following form

$$\frac{\partial^2 s}{\partial x'^2} - \frac{\partial^2 s}{\partial y'^2} + 4\frac{\partial^2}{\partial x'^2}\left(|u|^2\right) = 0, \tag{5.10}$$

$$i\frac{\partial u}{\partial t'} + \frac{\partial^2 u}{\partial x'^2} + \frac{\partial^2 u}{\partial y'^2} + 2|u|^2 u + s\,u = R[u], \tag{5.11}$$

where $x' = (k^2/\sqrt{\alpha_1})(c_g^{-1}x - t)$, $y' = k^2 y$, $t' = (\beta_1 k^4/\alpha_1)t$ and $R[u] = (1 - \kappa_1)\partial^2 u/\partial y'^2 + 2(1 - \kappa_2)|u|^2 u$ with $\kappa_1 = \alpha_1\beta_2/\beta_1$ and $\kappa_2 = 2\alpha_1\beta_3/(\alpha_2\beta_4)$.

For an arbitrary value of the wave number k, an exact 2D soliton solution decaying in all spatial directions is not available yet. But we notice that for small k one has $1 - \kappa_1 = -k^2/(6c^2) + k^4/(18c^4) - k^6/(54c^6) + O(k^8)$ and $1 - \kappa_2 = -k^2/(3c^2) - k^4/(18c^4) + k^6/(108c^6) + O(k^8)$, thus $R[u]$ is a small quantity proportional to k^2. In this case one can take $R[u]$ as a perturbation. We neglect $R[u]$ here. Then Eq. (5.11) is simplified as

$$i\frac{\partial u}{\partial t'} + \frac{\partial^2 u}{\partial x'^2} + \frac{\partial^2 u}{\partial y'^2} + 2|u|^2 u + s\,u = 0. \tag{5.12}$$

Equations (5.10) and (5.12) are standard type-I Davey-Stewartson (DSI) equations . They are completely integrable and can be solved exactly by the inverse scattering trans-

form. One of remarkable properties of the DSI equations is that they allow dromion
solutions decaying in all spatial directions [Ablowitz and Clakson (1991)].

The dromion solution of the DSI equations (5.10) and (5.12) reads

$$u = \frac{G}{F}, \quad s = 4\frac{\partial^2}{\partial x'^2}\ln F, \tag{5.13}$$

with

$$F = 1 + \exp(\eta_1 + \eta_1^*) + \exp(\eta_2 + \eta_2^*) + \gamma\exp(\eta_1 + \eta_1^* + \eta_2 + \eta_2^*), \tag{5.14}$$

$$G = \rho\exp(\eta_1 + \eta_2), \tag{5.15}$$

$$\eta_1 = (k_r + ik_i)x'' + (\Omega_r + i\Omega_i)\,t', \tag{5.16}$$

$$\eta_2 = (l_r + il_i)y'' + (\omega_r + i\omega_i)\,t', \tag{5.17}$$

$$\Omega_r = -2k_rk_i, \quad \omega_r = -2l_rl_i, \tag{5.18}$$

$$\Omega_i + \omega_i = k_r^2 + l_r^2 - k_i^2 - l_i^2, \tag{5.19}$$

$$\rho = |\rho|\exp(i\varphi_\rho), \quad |\rho| = 2[2k_rl_r(\gamma - 1)]^{1/2}, \tag{5.20}$$

where $k_r, k_i, l_r, l_i, |\rho|, \varphi_\rho$ and γ are real integerable constants. If choosing $k_rl_r > 0$ we
have $\gamma = \exp(2\varphi_\gamma)$, $\varphi_\gamma > 0$. x'' and y'' are the orthogonal transformation of x' and
y', i.e. $x'' = (x' + y')/\sqrt{2}$ and $y'' = (y' - x')/\sqrt{2}$. If taking $k_r = \sqrt{2}\sigma$, $l_r = \sqrt{2}\lambda$
($\lambda\sigma \geq 0$), $k_i = \sqrt{2}a$, $l_i = \sqrt{2}p$, $\Omega_i = 2(\sigma^2 - a^2)$, one has $\Omega_r = -4a\sigma$, $\omega_r = -4\lambda p$, and
$\omega_i = 2(\lambda^2 - p^2)$. Then we obtain

$$u = \frac{2\sigma\exp(ih)}{n_1\cosh f_1 + n_2\cosh f_2}, \tag{5.21}$$

$$s = \frac{4(n_1^2 + n_2^2)(\sigma^2 + \lambda^2) - 8\sigma^2}{(n_1\cosh f_1 + n_2\cosh f_2)^2}$$

$$+ \frac{8n_1n_2\left[(\sigma^2 + \lambda^2)\cosh f_1\cosh f_2 - (\sigma^2 - \lambda^2)\sinh f_1\sinh f_2\right]}{(n_1\cosh f_1 + n_2\cosh f_2)^2}, \tag{5.22}$$

where $n_1 = (\sigma/[\lambda(\gamma - 1)])^{1/2}$, $n_2 = (\sigma\gamma/[\lambda(\gamma - 1)])^{1/2}$ with $h = \sqrt{2}ax'' + \sqrt{2}py'' +$
$2(\sigma^2 + \lambda^2 - a^2 - p^2)t' + \varphi_\rho$, $f_1 = \sqrt{2}\sigma x'' - \sqrt{2}\lambda y'' - 4(a\sigma - \lambda p)t'$ and $f_2 = \sqrt{2}\sigma x'' +$
$\sqrt{2}\lambda y'' - 4(a\sigma + \lambda p)t' + \varphi_\gamma$. Obviously, the expression of u in (5.21) denotes a localized
envelope function *decaying in all spatial directions*, called dromion [Ablowitz and Clak-
son (1991)]. From (5.21) we know that the dromion has an amplitude $2\sigma/(n_1^2 + n_2^2)^{1/2}$
and, in the $Ox''y''$ coordinate system, at time t' it locates at the position $(x'', y'') =$
$(4at'/\sqrt{2} - \varphi_\gamma/(2\sqrt{2}\sigma), 4pt'/\sqrt{2} - \varphi_\gamma/(2\sqrt{2}\lambda))$. Hence the dromion has a constant velocity $\mathbf{V}_d = (4a/\sqrt{2}, 4p/\sqrt{2})$.

The mean field component s consists of two interacting plane solitons with each plane
soliton decaying in its travelling direction. It is easy to show that s has the following
asymptotic form $s|_{x''\to-\infty} = 4\lambda^2\,\text{sech}^2\,\lambda(\sqrt{2}y'' - 4pt')$, $s|_{x''\to+\infty} = 4\lambda^2\text{sech}^2\,[\lambda(\sqrt{2}y'' - 4pt'$
$s|_{y''\to-\infty} = 4\sigma^2\text{sech}^2\,\sigma(\sqrt{2}x'' - 4at')$, and $s|_{y''\to+\infty} = 4\sigma^2\text{sech}^2\,[\sigma(\sqrt{2}x'' - 4at') + \varphi_\gamma]$. Thus
in the $Ox''y''$ coordinate system the plane soliton with the amplitude $4\lambda^2$ (λ-soliton) trav-
els with the velocity $\mathbf{V}_\lambda = (0, 4p/\sqrt{2})$, while the plane soliton with the amplitude $4\sigma^2$

(σ-soliton) travels with the velocity $\mathbf{V}_\sigma = (4a/\sqrt{2}, 0)$. There is a overlapping region (corresponding to an oblique collision between the plane solitons) where a new plane soliton, i. e. a Mach stem, appears. The center-point of the overlapping region of two plane solitons is just the position of the dromion and its velocity is also $\mathbf{V}_d$. Thus *the dromion rides exactly on the cross-point of the two plane solitons and travels with the common velocity $\mathbf{V}_d$.* From this result we see that the dromion, which represents the high-frequency component of the excitation, can be taken as being driven by the "truck", i. e. the long wavelength (low-frequency) component denoted by two plane solitons.

Now we give the explicit expression for the order parameter when the dromion excitation presented above is created. At the leading order approximation we get

$$\psi = P(x, y, t) \exp\left(-i\mu t - \frac{1}{2}z^2 + i\varphi(x, y, t)\right) \tag{5.23}$$

where

$$P(x, y, t) = u_0\left(1 - B_0 \frac{\sin \Phi}{n_1 \cosh f_1 + n_2 \cosh f_2}\right), \tag{5.24}$$

$$\varphi = -\frac{\beta_1}{\sqrt{\alpha_1 \beta_4}} k^2 D_0(x, y, t) + \left(\frac{4\beta_1}{\alpha_2 \beta_4}\right)^{1/2} k^4 \frac{4\sigma \cos \Phi}{n_1 \cosh f_1 + n_2 \cosh f_2}, \tag{5.25}$$

with

$$\Phi = \left[k + (a - p)\frac{k^2}{\sqrt{\alpha_1}}\right] x + (a + p)k^2 y$$

$$+ \left[2(\sigma^2 + \lambda^2 - a^2 - p^2)\frac{\beta_1}{\alpha_1} k^4 - (a - p)c_g \frac{k^2}{\sqrt{\alpha_1}}\right] t + \varphi_\rho, \tag{5.26}$$

$$f_1 = (\sigma + \lambda)\frac{k^2}{\sqrt{\alpha_1}} x + (\sigma - \lambda)k^2 y - \left[(\sigma + \lambda)c_g \frac{k^2}{\sqrt{\alpha_1}} + 4(a\sigma - \lambda p)\frac{\beta_1}{\alpha_1} k^4\right] t, \tag{5.27}$$

$$f_2 = (\sigma - \lambda)\frac{k^2}{\sqrt{\alpha_1}} x + (\sigma + \lambda)k^2 y - \left[(\sigma - \lambda)c_g \frac{k^2}{\sqrt{\alpha_1}} + 4(a\sigma + \lambda p)\frac{\beta_1}{\alpha_1} k^4\right] t + \tag{5.28}$$

$$B_0 = 4\sigma \left(\frac{4\beta_1}{\alpha_2 \beta_4}\right)^{1/2} \frac{k^5}{\sqrt{k^2 + 4c^2}}, \tag{5.29}$$

$$D_0 = \frac{\sigma \exp(\eta_1 + \eta_1^*)\left[1 + \gamma \exp(\eta_2 + \eta_2^*)\right] + \lambda \exp(\eta_2 + \eta_2^*)\left[1 + \gamma \exp(\eta_1 + \eta_1^*)\right]}{1 + \exp(\eta_1 + \eta_1^*) + \exp(\eta_2 + \eta_2^*) + \gamma \exp(\eta_1 + \eta_1^* + \eta_2 + \eta_2^*)} \tag{5.30}$$

with

$$\eta_1 + \eta_1^* = 2\sigma\left\{\frac{k^2}{\sqrt{\alpha_1}} x + k^2 y - \left[\frac{k^2}{\sqrt{\alpha_1}} c_g + 4a\frac{\beta_1}{\alpha_1} k^4\right] t\right\}, \tag{5.31}$$

$$\eta_2 + \eta_2^* = 2\lambda\left\{-\frac{k^2}{\sqrt{\alpha_1}} x + k^2 y + \left[\frac{k^2}{\sqrt{\alpha_1}} c_g - 4p\frac{\beta_1}{\alpha_1} k^4\right] t\right\}. \tag{5.32}$$

From the expression (5.24) we see that the excitation is a *gray dromion* created from the background (the ground-state condensate). The parameter B_0 characterizes the its *grayness*. The phase correction of the order parameter, i. e. φ given by (5.25), includes two parts. The first part is a mean flow (represented by D_0) describing an oblique interaction of two plane kinks. The second part is a dromion decaying exponentially in all spatial directions.

Figure 6 shows the modulus and phase of the solution given by Eqs. (5.24) and (5.25).

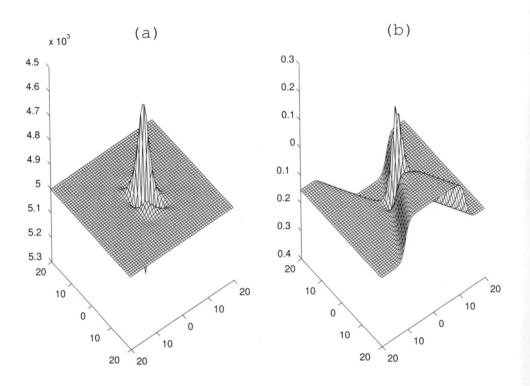

Figure 6. Dromion excitation in the BEC. (a) and (b) are the modulus P and the phas e of the order parameter ψ, respectively. The parameters are choosen as $\sigma = \lambda = a = p = 1$, $\varphi_\rho = 0$, $\varphi_\gamma = 1$, $k = 0.6$, and $t = 1.0$.

6 Second-harmonic generation in a two-component BEC

Wave resonant interaction is a classical topic in nonlinear optics [Shen (1984)]. It also occurs widely in fluid mechanics and plasma physics [Craik (1985)]. It is natural to look for the possibility of such interaction of matter waves. In 1999, Deng et al demonstrated successfully a four-wave mixing (FWM) by using phase-matched BEC wave packets [Deng et al (1999)]. A FWM is allowed in BECs because the nonlinearity in the GP equation (2.7) is cubic. A parametric resonant interaction, which requires a quadratic nonlinearity in the equation of motion, seems not possible in BECs. However, if we consider the excitations created from a BEC background, the interaction between the excitations is of quadratic nonlinearity and hence the parametric interaction can occur.

In this section we investigate the second harmonic generation (SHG) of propagating collective excitations in a two-component BEC. Note that for a single-component BEC the SHG of excitation can not happen because the excitation spectrum in this case [see (3.6)] can not fulfill the phase-matching condition for the SHG. Thus we turn to consider a two-component BEC, whose excitation spectrum displays two branches and hence provides us with the possibility to fulfill the SHG phase-matching condition (see below). Since the SHG is a process of energy up-conversion, at zero temperature such process can be well described by two coupled Gross-Pitaevskii (GP) equations. Using a method of multiple-scales we derive the nonlinearly coupled amplitude equations describing the SHG and give their explicit solutions. We show that an experimental realization of such SHG may give information about the interaction between different components of the condensate.

The two-component BEC we are considering may be a relatively arbitrary binary mixture of two condensates. Such mixture may consist of different particles such as ^{87}Rb and ^{23}Na, or different isotopes such as ^{87}Rb and ^{85}Rb, or different hyperfine spin states of the same species. Denoting $\psi_j(\mathbf{r},t)$ as the order parameter of species j with particle number $N_j = \int d\mathbf{r}|\psi_j|^2$ $(j=1,2)$, ψ_j satisfy the coupled GP equations

$$i\hbar\frac{\partial\psi_1}{\partial t} = \left[-\frac{\hbar^2}{2m_1}\nabla^2 + V_1(\mathbf{r}) + g_{11}|\psi_1|^2 + g_{12}|\psi_2|^2\right]\psi_1, \tag{6.1}$$

$$i\hbar\frac{\partial\psi_2}{\partial t} = \left[-\frac{\hbar^2}{2m_2}\nabla^2 + V_2(\mathbf{r}) + g_{21}|\psi_1|^2 + g_{22}|\psi_2|^2\right]\psi_2, \tag{6.2}$$

where m_j and $V_j(\mathbf{r})$ are respectively the atomic mass and external trapping potential for the species j, $g_{jl} = 2\pi\hbar^2 a_{jl}/m_{jl}$ is interaction parameter with a_{jl} $(j,l = 1,2)$ being the s-wave scattering length between the species j and the species l $(a_{jl} > 0$ for repulsive interaction) and $m_{jl}=m_j m_l/(m_j+m_l)$ being the reduced mass. We consider a quasi one-dimensional (1D) trap with a negligible axial confinement. Thus the trapping potentials take the form $V_j(\mathbf{r})=(m_j/2)\omega_{j\perp}^2(y^2 + z^2)]$, where $\omega_{j\perp}$ are the trap frequencies of the species j in transverse directions. For simplicity we assume $m_1=m_2=m$ and $\omega_{1\perp}=\omega_{2\perp}=\omega_\perp$. Such assumption is only for getting simplified expressions and clarifying essential physics. A more general case can be considered with a similar result given below.

As in the last three sections by expressing the order parameters in terms of their modulus and phases, i. e. $\psi_j=\sqrt{n_j}\exp(i\phi_j)$, we obtain a set of coupled nonlinear equations for n_j and ϕ_j $(j=1,2)$. By letting the condensate densities, time, axial spatial coordinate and transverse spatial coordinates are measured respectively in the units $n_0=N_1/(La_\perp^2)$, $t_0=\omega_\perp^{-1}$, L (condensate length in the axial direction), and $a_\perp=[\hbar/(m\omega_\perp)]^{1/2}$ (harmonic oscillator length in the transverse directions), these equations become dimensionless. Assuming that the transverse confinement is strong one can take $\sqrt{n_j} = A_j(x,t)\pi^{-1/2}$

$\exp[-(y^2 + z^2)/2]$ and $\phi_j = \phi_j(x, t)$ $(j = 1, 2)$. Then we obtain

$$\frac{\partial A_j}{\partial t} + \varepsilon \left[\frac{\partial A_j}{\partial x} \frac{\partial \phi_j}{\partial x} + \frac{1}{2} A_j \frac{\partial^2 \phi_j}{\partial x^2} \right] = 0, \tag{6.3}$$

$$\left(\frac{\partial \phi_j}{\partial t} + 1 \right) A_j$$
$$+ \varepsilon \left[-\frac{1}{2} \frac{\partial^2}{\partial x^2} + \frac{1}{2} \left(\frac{\partial \phi_j}{\partial x} \right)^2 + \frac{1}{2\pi} \left(G_{jj} A_j^2 + G_{j\,3-j} A_{3-j}^2 \right) \right] A_j = 0, \tag{6.4}$$

with $G_{ij} = g_{ij}/g_{11}$. The parameter $\varepsilon = n_0 g_{11}/(\hbar\omega_\perp)$, i. e. the ratio between the atomic interaction and the strength of the transverse confinement.

The stationary (ground) state solution of the above equations is given by $A_1 = A_{1GS} = 1/\sqrt{L}$, $A_2 = A_{2GS} = \sqrt{N_2/(N_1 L)}$, $\phi_{1GS} = -(1 + \mu_1^{(1)})t$, and $\phi_{2GS} = -(1 + \mu_2^{(1)})t$ with $\mu_1^{(1)} = (G_{11} + G_{12} N_2/N_1)/(2\pi L)$, and $\mu_2^{(1)} = (G_{21} + G_{22} N_2/N_1)/(2\pi L)$. The linear dispersion relation of an excitation from the ground state is given by

$$\omega^2(q)/q^2 = \frac{1}{2} \left(\tilde{G}_{11} + \tilde{G}_{22} \right) + \frac{1}{4} q^2 \pm \frac{1}{2} \left[\left(\tilde{G}_{11} - \tilde{G}_{22} \right)^2 + 4\tilde{G}_{12} \tilde{G}_{21} \right]^{1/2}, \tag{6.5}$$

where $\tilde{G}_{11} = G_{11}/(2\pi L)$, $\tilde{G}_{12} = G_{12} N_2/(2\pi L N_1)$, $\tilde{G}_{21} = G_{21}/(2\pi L)$, and $\tilde{G}_{22} = G_{22} N_2/(2\pi L N_1)$. q and ω are wavevector and frequency of the excitation, respectively. From (6.5) we see that dispersion curve of the collective modes has two branches, i. e. the upper branch $\omega_+(q)$ and the lower branch $\omega_-(q)$, both of them are acoustic.

A necessary condition for the SHG is that the phase-matching condition, i. e.

$$q_2 = 2q_1, \qquad \omega_2 = 2\omega_1, \tag{6.6}$$

must be fulfilled, where q_1 (q_2) and ω_1 (ω_2) are the wave vector and frequency of the fundamental (second harmonic) wave, respectively. By choosing $\omega_1 = \omega_+(q_1)$ and $\omega_2 = \omega_-(q_2) = \omega_-(2q_1)$, the condition (6.6) is equivalent to $\omega_-(2q_1) = 2\omega_+(q_1)$, which results in the solution

$$q_1 = \frac{2}{\sqrt{3}} \left[\left(\tilde{G}_{11} - \tilde{G}_{22} \right)^2 + 4\tilde{G}_{12} \tilde{G}_{21} \right]^{1/4}. \tag{6.7}$$

We see that for the two-component condensate the SHG phase-matching condition can be easily satisfied. Shown in Fig. 7 is the dispersion curve of the collective modes of the two-component BEC consisting of different hyperfine spin states. The modes satisfying the phase-matching condition (6.6) have been clearly shown as the point $A = (q_1, \omega_1)$ (the fundamental wave) and the point $B = (q_2, \omega_2)$ (the second harmonic wave).

To derive the amplitude equations for the SHG we make the asymptotic expansion $A_j - A_{jGS} = A_{jGS}(\varepsilon F_j^{(1)} + \varepsilon^2 F_j^{(2)} + \cdots)$ and $\phi_j - \phi_{jGS} = \varepsilon \phi_j^{(1)} + \varepsilon^2 \phi_j^{(2)} + \cdots$, where $F_j^{(l)}$ and $\phi_j^{(l)}$ are the functions of the fast variables x, τ and the slow variables $X = \varepsilon x$, $T = \varepsilon \tau$. The expansion parameter $\varepsilon [= n_0 g_{11}/(\hbar\omega_\perp)]$ can be small as long as the typical

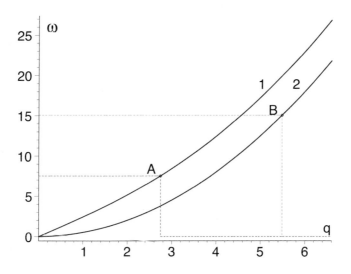

Figure 7. The dispersion curve of a two-component BEC consisting of two different hyperfine spin states. Curves 1 and 2 represent the upper branch, $w_+(q)$, and the lower branch, $w_-(q)$, respectively. For the fundamental wave and the second harmonic wave, the phase-matched wave vectors and frequencies for the SHG are ($q_1 = 2.743, w_1 = 7.536$) and ($q_2 = 5.486, w_2 = 15.072$), which have been illustrated by the points $A = (q_1, w_1)$ and $B = (q_2, w_2)$, respectively.

value of the chemical potential is less than the level spacing of the harmonic oscillator. Using such expansion Eqs.(6.3)-(6.4) are transferred into a set of equations for $F_j^{(l)}$ and $\phi_j^{(l)}$ ($j = 1, 2; l = 1, 2, 3, ...$).

In the first-order ($l = 1$) we get the solution in linear approximation. For the SHG we take

$$F_1^{(1)} = U_1 \exp(i\theta_1) + U_2 \exp(i\theta_2) + c.c., \tag{6.8}$$

$$F_2^{(1)} = \left(\tilde{G}_{12}q_1^2\right)^{-1} L_1(w_1, q_1)U_1 \exp(i\theta_1)$$

$$+ \left(\tilde{G}_{12}q_2^2\right)^{-1} L_1(w_2, q_2)U_2 \exp(i\theta_2) + c.c., \tag{6.9}$$

$$\phi_1^{(1)} = -i(2w_1/q_1^2)U_1 \exp(i\theta_1) - i(2w_2/q_2^2)U_2 \exp(i\theta_2) + c.c., \tag{6.10}$$

$$\phi_2^{(1)} = -i2\gamma_m w_1 \left(\tilde{G}_{12}q_1^4\right)^{-1} L_1(w_1, q_1)U_1 \exp(i\theta_1)$$

$$- i2\gamma_m w_2 \left(\tilde{G}_{12}q_2^4\right)^{-1} L_1(w_2, q_2)U_2 \exp(i\theta_2) + c.c., \tag{6.11}$$

where $L_1(w, q) = w^2 - (\tilde{G}_{11} + q^2/4)q^2$, U_1 and U_2 are respectively the envelope functions of the fundamental wave (with the phase $\theta_1 = q_1 x - w_1 \tau$) and the second harmonic wave (with the phase $\theta_2 = q_2 x - w_2 \tau$). q_1, q_2, w_1 and w_2 are chosen according to the SHG phase-matching condition (6.6), i.e. $w_1 = w_+(q_1)$ and $w_2 = w_-(q_2)$ with $q_2 = 2q_1$.

At the second-order ($l = 2$), solvability conditions give the closed equations for U_1 and U_2. After making the transformation $U_j = \varepsilon u_j$ and noting that $X = \varepsilon x$ and $T = \varepsilon \tau$, we get the amplitude equations:

$$\frac{\partial u_1}{\partial \tau} + v_{g1} \frac{\partial u_1}{\partial x} + i\Gamma_1 u_1^* u_2 \exp(-i\Delta q x) = 0, \tag{6.12}$$

$$\frac{\partial u_2}{\partial \tau} + v_{g2} \frac{\partial u_2}{\partial x} + i\Gamma_2 u_1^2 \exp(-i\Delta q x) = 0, \tag{6.13}$$

where $v_{gj} = (d\omega_+/dq)_{q=q_j}$ is the group velocity of $j$the wave, $\Delta q = q_2 - 2q_1$ is a possible phase mismatch. The explicit expressions of v_{gj} and Γ_j ($j = 1, 2$) have been given in Huang et al (2004).

The solutions of Eqs. (6.12) and (6.13) have been given in nonlinear optics and fluid mechanics [Shen (1984), Craik (1985)]. For a stationary case ($\partial/\partial\tau = 0$) and for $\Delta q = 0$, Eqs. (6.12) and (6.13) admit the solution:

$$u_1 = \left(\frac{-\Gamma_1 W}{v_{g1}} \right)^{1/2} \mathrm{sech} \left[\frac{\Gamma_1}{v_{g1}} \left(\frac{-\Gamma_2 W}{v_{g2}} \right)^{1/2} x \right] e^{i\varphi_0}, \tag{6.14}$$

$$u_2 = \left(\frac{-\Gamma_2 W}{v_{g2}} \right)^{1/2} \tanh \left[\frac{\Gamma_1}{v_{g1}} \left(\frac{-\Gamma_2 W}{v_{g2}} \right)^{1/2} x \right] e^{i(2\varphi_0 + \pi/2)}, \tag{6.15}$$

where $W(x) = -(v_{g1}/\Gamma_1)|u_1|^2 - (v_{g2}/\Gamma_2)|u_2|^2 = W(0)$ is a constant (denoting the input power of the excitation) and φ_0 is an arbitrary constant. At $x = 0$, the fundamental wave takes the total power W of the system and thus power of the second harmonic wave is zero. As x increases the power of the fundamental wave is converted gradually into the second harmonic wave. At distance x, the conversion efficiency from the fundamental wave into the second harmonic wave is given by

$$\eta = \frac{W_2(x)}{W_1(0)} = \tanh^2 \left[\frac{\Gamma_1}{v_{g1}} \left(\frac{-\Gamma_2 W}{v_{g2}} \right)^{1/2} x \right], \tag{6.16}$$

where $W_j(x) = -(v_{gj}/\Gamma_j)|u_j|^2$ is the power of jth wave. Thus the conversion efficiency of the SHG is determined by v_{gj}, Γ_j ($j = 1, 2$), W and the propagating distance x. Since v_{gj} and Γ_j depends on the interatomic interaction parameters g_{ij}, a larger conversion efficiency can be obtained by controlling g_{ij}.

Shown in Fig. 8 is the conversion efficiency η as a function of G_{12} ($= G_{21}$) and the propagating distance (or sample length) x. The curves 1, 2, and 3 correspond to the propagating distance x taking the values 5.0, 10.0 and 20.0, respectively. From Fig. 8 we see that to obtain a significant conversion efficiency of the SHG, in addition to a larger propagating distance and a larger input power, one must choose an appropriate interspecies interaction strength G_{12}. Inversely, it also provides us with a possibility for determining G_{12} by measuring the SHG conversion efficiency η.

For very short-pulse excitations the walk-off effect due to different group velocities for the fundamental and the second harmonic wave must be taken into account [Shen (1984)]. By the transformation $u_1 = [v_{g1}v_{g2}/(\Gamma_1\Gamma_2)]^{1/2} w_1 \exp(i\varphi)$ and $u_2 = $

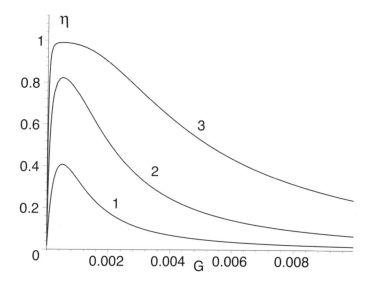

Figure 8. The conversion efficiency as a function of interspecies interaction parameter G $(= G_{12} = G_{21})$ and propagating distance x in a stationary SHG for $G_{11} = 1.0$, $G_{22} = 1.0027$. Curves 1, 2, and 3 correspond to the input power $W = 10.0$ with the propagating distance x taking the values 5.0, 10.0 and 20.0, respectively.

$(v_{g1}/\Gamma_1)w_2 \exp[i(2\varphi + \pi/2)]$ (w_1, w_2 and φ are real functions of x and τ) and assumption $\Delta q = 0$, Eqs. (6.12) and (6.13) become

$$\frac{\partial w_1}{\partial x} + \frac{1}{v_{g1}} \frac{\partial w_1}{\partial \tau} = w_1 w_2, \tag{6.17}$$

$$\frac{\partial w_2}{\partial x} + \frac{1}{v_{g2}} \frac{\partial w_2}{\partial \tau} = -w_1^2. \tag{6.18}$$

Consider a travelling-wave solution, i. e. take w_j $(j = 1, 2)$ are the functions of x and $\zeta = \tau - x/v_{g1}$, Eqs. (6.17) and (6.18) are transferred as $\partial w_1/\partial x = w_1 w_2$, $\partial w_2/\partial x + \nu \partial w_2/\partial \zeta = -w_1^2$, where $\nu = 1/v_{g2} - 1/v_{g1}$ is a parameter denoting the walk-off (or group-velocity dispersion) effect. If at $x = 0$ the fundamental wave and the second harmonic wave take the form $w_1(0, \tau) = A_0/(1 + \tau^2/\tau_0^2)$ and $w_2(0, \tau) = 0$, where τ_0 is the initial pulse-width and A_0 is a constant representing the initial amplitude of the fundamental wave, one can get the following solution [Shen (1984)]:

$$w_1(\zeta, x) = \frac{A_0}{\left(1 + \tilde{\zeta}^2\right)^{1/2} \left[1 + \left(\tilde{\zeta} - \tilde{x}\right)^2\right]^{1/2}} \frac{1}{\cosh \xi + \tilde{\zeta}/f_0 \sinh \xi}, \tag{6.19}$$

$$w_2(\zeta, x) = \frac{\tau_{cr}}{\tau} \frac{A_0}{1 + \left(\tilde{\zeta} - \tilde{x}\right)^2} \frac{\tilde{x} \cosh \xi + \left[f_0 - \tilde{\zeta}(\tilde{\zeta} - \tilde{x})/f_0\right] \sinh \xi}{\cosh \xi + \tilde{\zeta}/f_0}, \tag{6.20}$$

where $\tilde{\zeta} = \zeta/\tau_0$, $\tilde{x} = x/L_\nu$, $f_0 = (\tau_0^2/\tau_{cr}^2 - 1)^{1/2}$, $\tau_{cr} = \nu L_{NL}$, $\xi = f_0[\tan^{-1}\tilde{\zeta} - \tan^{-1}(\tilde{\zeta} - \tilde{x})]$ with $L_\nu = \tau_0/\nu$ (walk-off or dispersion length) and $L_{NL} = A_0^{-1}$ (nonlinear length). If the walk-off length is much larger than the nonlinear length, i. e. $L_\nu >> L_{NL}$, the walk-off effect can be neglected. In this case the solution (6.19) and (6.20) is simplified into $w_1(\zeta, x) = A_0/(1 + \tilde{\zeta}^2) \operatorname{sech}\left[A_0 x/(1 + \tilde{\zeta})\right]$ and $w_2(\zeta, x) = A_0/(1 + \tilde{\zeta}^2) \tanh\left[A_0 x/(1 + \tilde{\zeta})\right]$. This situation corresponds to a quasi-stationary SHG and only in this case the conversion efficiency is significant.

7 Summary

A Bose-condensed gas is a coherent quantum many-body system which at zero temperature can be described by a classical nonlinear wave equation (i. e. the GP equation). A BEC with a repulsive interparticle interaction can display very rich soliton excitations. We have shown that a long wavelength, weak nonlinear excitation can be described by the KdV equation for a cigar-shaped BEC and by the KP equation for a disk-shaped BEC. The nonlinear excitations with a short wavelength in a disk-shaped BEC obey the Davey-Stewartson equations. Corresponding soliton solutions have been provided and their physical properties have been discussed. We have also shown that a second harmonic generation for the nonlinear excitations in a two-component BEC is possible.

If the interparticle interaction is attractive, one can show that the soliton excitations in BECs are bright ones. A lot of theoretical and experimental studies exist on this topic. For detail, see Meystre (2001), Strecker et al (2002), and Khaykovich et al (2002).

Acknowledgments

The author thanks M. G. Velarde, V. A. Makarov, and J. Szeftel for fruitful discussions and collaboration. This research was supported by the Natural Science Foundation of China under grant.

Bibliography

M. J. Ablowitz and P. A. Clarkson. *Solitons, Nonlinear Evolution Equations and Inverse Scattering*. London Mathematical Society Lecture Note Series 149, Camb. Univ. Press, 1991.

G. P. Agrawal. *Nonlinear Fiber Optics*, 3rd ed. Academic, New York, 2001.

M. H. Anderson, J. R. Ensher, M. R. Matthews, C. E. Wieman, and E. A. Cornell. Observation of Bose-Einstein condensation in a dilute atomic vapor. *Science* 269: 198, 1995.

M. R. Andrews, D. M. Kurn, H.-J. Miesner, D. S. Durfee, C. G. Townsend, S. Inouye, and W. Ketterle. Propagation of sound in a Bose-Einstein condensate. *Phys. Rev. Lett.* 79: 553, 1998.

C. C. Bradley, C. A. Sackett, J. J. Tollett, and R. G. Hulet. Evidence of Bose-Einstein condensation in an atomic gas with attractive interactions. *Phys. Rev. Lett.* 75: 1687, 1995.

S. Burger, K. Bongs, S. Dettmer, W. Ertmer, and K. Sengstock. Dark solitons in Bose-Einstein condensates. *Phys. Rev. Lett.* 83: 5198, 1999.

S. Burger, L. D. Carr, P. Ohberg, K. Sengstock, and A. Sanpera. Generation and interaction of solitons in Bose-Einstein condensates. *Phys. Rev. A* 65: 043611, 2002; J. Brand and W. P. Reinhardt. Solitonic vortices and the fundamental modes of the snake instability: Possibility of observation in the gaseous Bose-Einstein condensate. *Phys. Rev. A* 65: 043612, 2002.

Th. Busch and J. R. Anglin, Motion of dark solitons in trapped Bose-Einstein condensates. *Phys. Rev. Lett.* 84: 2298, 2000; D. L. Feder, M. S. Pindzola, L. A. Collins, B. I. Schneider, and C. W. Clark. Dark-soliton states of Bose-Einstein condensates in anisotropic traps. *Phys. Rev. A* 62: 053606, 2000.

A. D. D. Craik, *Wave Interactions and Fluid Flows*. Cambridge Univ. Press, Cambridge, 1985.

W. Cui, C. Sun, and G. Huang. Dromion excitations in self-defocusing optical media. *Chinese Phy. Lett.* 20: 246, 2003.

F. Dalfovo, S. Giorgini, L. P. Pitaevskii, and S. Stringari. Theory of Bose-Einstein condensation in trapped gases. *Rev. Mod. Phys.* 71: 463, 1999.

A. Davey and K. Stewartson. On three-dimensional packets of surface wave. *Proc. R. Soc. London Ser. A* 338: 101, 1974.

K. B. Davis, M.-O. Mewes, M. R. Andrews, N. J. van Druten, D. S. Durfee, D. M. Kurn, and W. Ketterle. Bose-Einstein condensation in a gas of sodium atoms. *Phys. Rev. Lett.* 75: 3969, 1995.

L. Deng, E. W. Hagley, J. Win, M. Trippenbach, Y. Band, P. S. Julienne, J. E. Simsarian, K. Helmerson, S. L. Rolston, and W. D. Phillips. Four-wave mixing with matter waves. *Nature* 398: 218, 1999.

J. Denschlag, J. E. Simsarian, D. L. Feder, C. W. Clark, L. A. Collins, J. Cubizolles, L. Deng, E. W. Hagley, K. Helmerson, W. P. Reinhart, S. L. Rolston, B. I. Schneider, and W. D. Phillips. Generating solitons by phase engineering of a Bose-Einstein condensate. *Science* 287: 97, 2000.

P. G. Drazin and R. S. Johnson. *Solitons: an Introduction*. Cambridge Univ. Press, Cambridge, 1989.

A. Einstein. Quantentheorie des einatomigen idealen gases: Zweite abhandlung. *Sitzungsbr. Klg. Preuss. Akad. Wiss.* 3, 1925.

E. V. Goldstein and P. Meystre, Phase conjugation of multicomponent Bose-Einstein condensates. *Phys. Rev. A* 59: 1509, 1999; K. Rzazewski, M. Trippenbach, S. J.

Singer, and Y. B. Band. Statistics of atomic populations in output coupled wave packets from Bose-Einstein condensates: Four-wave mixing. *Phys. Rev. A* 61: 013606, 1999; M. G. Moore and P. Meystre. Theory of superradiant scattering of laser light from Bose-Einstein condensates. *Phys. Rev. Lett.* 83: 5202, 1999; Y. Wu, X. Yang, C. P. Sun, X. J. Zhou, and Y. Q. Wang. Theory of four-wave mixing with matter waves without the undepleted pump approximation. *Phys. Rev. A* 61: 043604, 2000; M. Trippenbach, Y. B. Band, and P. S. Julienne. Theory of four-wave mixing of matter waves from a Bose-Einstein condensate. *Phys. Rev. A* 62: 023608, 2000.

A. Görlitz, J. M. Vogels, A. E. Leanhardt, C. Raman, T. L. Gustavson, J. R. Abo-Shaeer, A. P. Chikkatur, S. Gupta, S. Inouye, T. Rosenband, and W. Ketterle. Realization of Bose-Einstein condensates in lower dimensions. *Phys. Rev. Lett.* 87: 130402, 2001.

M. Greiner, C. A. Regal, and D. S. Jin. Emergence of a molecular Bose-Einstein condensate from a Fermi gas. *Nature* 426: 537, 2003.

R. Grimshaw. *Korteweg-de Vries Equation*, this volume.

G. Huang. KdV description of solitons in Bose-Einstein condensation. *Chinese Phys. Lett.* 18: 628, 2001.

G. Huang, M. G. Velarde, and V. A. Makarov. Dark solitons and their head-on collisions in Bose-Einstein condensates. *Phys. Rev. A* 64: 013617, 2001a.

G. Huang, V. V. Konotop, and M. G. Velarde. Nonlinear modulation of multidimensional lattice waves. *Phys. Rev. E* 64: 056619, 2001b.

G. Huang, J. Szeftel, and S. Zhu, Dynamics of dark solitons in quasi-one-dimensional Bose-Einstein condensates. *Phys. Rev. A* 65: 053605, 2002.

G. Huang, V. A. Makarov, and M. G. Velarde. Two-dimensional solitons in Bose-Einstein condensates with a disk-shaped trap. *Phys. Rev. A* 67: 023604, 2003.

G. Huang, X.-q. Li, and J. Szeftel. Second-harmonic generation of Bogoliubov excitations in a two-component Bose-Einstein condensate. *Phys. Rev. A* 69: 065601, 2004.

S. Inouye, T. Pfau, S. Gupta, A. P. Chikkatur, A. Görlitz, D. E. Pritchard, and W. Ketterle. Phase-coherent amplification of atomic matter waves. *Nature* (London) 402: 641, 1999a.

S. Inouye, A. P. Chikkatur, D. M. Stamper-Kurn, J. Stenger, D. E. Pritchard, and W. Ketterle. Superradiant Rayleigh scattering from a Bose-Einstein condensate. *Science* 23: 571, 1999b.

D. S. Jin, J. R. Ensher, M. R. Matthews, C. E. Wieman, and E. A. Cornell D. S. Jin et al. Collective excitations of a Bose-Einstein condensate in a dilute gas. *Phys. Rev. Lett.* 77: 420, 1996.

S. Jochim, M. Bartenstein, A. Altmeyer, G. Hendl, S. Riedl, C. Chin, J. Hecker, J. Denschlag, and R. Grimm, Bose-Einstein condensation of molecules. *Science* 302: 2101, 2003.

L. Khaykovich, F. Schreck, G. Ferrari, T. Bourdel, J. Cubizolles, L. D. Carr, Y. Castin, and C. Salomon. Formation of a matter-wave bright soliton. *Science* 296: 1290, 2002.

D. B. Khismatulin and I. Sh. Akhatov. Sound-ultrasound interaction in bubbly fluids: Theory and possible applications. *Phys. Fluids* 13: 3582, 2001.

M. Kozuma, Y. Suzuki, Y. Torii, T. Sugiura, T. Kuga, E. W. Hagley, and L. Deng Phase-coherent amplification of matter waves. *Science* 286: 2309, 1999.

A. J. Legget, Bose-Einstein condensation in the alkali gases: Some fundamental concepts. *Rev. Mod. Phys.* 73: 307, 2001.

H. Linde, X.-L. Chu, and M. G. Velarde. Oblique and head-on collisions of solitary waves in Marangoni-Bènard convection. *Phys. Fluids A* 5: 1068, 1993.

Mattews, M. R., B. P. Anderson, P. C. Haljan, D. S. Hall, C. E. Wieman, and E. A. Cornell. Vortices in a Bose-Einstein condensate. *Phys. Rev. Lett.* 83: 2498, 1999; K. W. Madison, F. Chevy, W. Wohlleben, and J. Dalibard. Vortex formation in a stirred Bose-Einstein condensate. *Phys. Rev. Lett.* 84: 806, 2000; J. R. Abo-Shaeer, C. Raman, J. M. Vogels, W. Ketterle. Observation of vortex lattices in Bose-Einstein condensates. *Science* 292: 476, 2001.

P. Meystre. *Atom Optics.* Springer-Verlag, New York, 2001.

M. G. Moore and P. Meystre. Atomic four-wave mixing: fermions versus bosons. *Phys. Rev. Lett.* 86: 4199, 2001; W. Ketterle and S. Inouye. Does matter wave amplification work for fermions?. *Phys. Rev. Lett.* 86: 4203, 2001.

A. C. Newell and J. V. Moloney, *Nonlinear Optics.* Addison Wesley, Red Wood, CA, 1992.

V. M. Perez-Garcia, H. Michinel, and H. Herrero. Bose-Einstein solitons in highly asymmetric traps. *Phys. Rev. A* 57: 3837, 1998; A. D. Jackson, G. M. Kavoulakis, and C. J. Pethick. Solitary waves in clouds of Bose-Einstein condensed atoms. *Phys. Rev. A* 58: 2417, 1998; O. Zobay, S. Pötting, P. Meystre, and E. M. Wright. Creation of gap solitons in Bose-Einstein condensates. *Phys. Rev. A* 59: 643, 1999; A. E. Muryshev, H. B. van Linden van den Heuvell, and G. V. Shlyapnikov. Stability of standing matter waves in a trap. *Phys. Rev. A* 60: R2665, 1999.

C. J. Pethick and H. Smith. *Bose-Einstein Condensation in Dilute Gases.* (Cambridge University Press, Cambridge, 2002).

S. L. Rolston and W. D. Phillips, Nonlinear and quantum atom optics. *Nature* 416: 219, 2002.

Y. R. Shen, *The Principles of Nonlinear Optics.* Wiley, New York, 1984.

K. E. Strecker, G. B. Partridge, A. G. Truscott, and R. G. Hulet. Formation and propagation of matter-wave soliton trains. *Nature* (London) 417: 150, 2002.

C. H. Su and R. M. Mirie. On head-on collisions between two solitary waves. *J. Fluid Mech.* 98: 509, 1980.

A. Trombettoni and A. Smerzi, Discrete solitons and breathers with dilute Bose-Einstein condensates. *Phys. Rev. Lett.* 86: 2353, 2001; P. Ohberg and L. Santos, Dark solitons in a two-component Bose-Einstein condensate. *Phys. Rev. Lett.* 86: 2918, 2001; B. P. Anderson, P. C. Haljan, C. A. Regal, D. L. Feder, L. A. Collins, C. W. Clark, and E. A. Cornell. Watching dark solitons decay into vortex rings in a Bose-Einstein condensate. *Phys. Rev. Lett.* 86: 2926, 2001; P. D. Drummond, A. Eleftheriou, K. Huang, and K. V. Kheruntsyan. Theory of a mode-locked atom laser with toroidal geometry. *Phys. Rev. A* 63: 053602, 2001.

N. J. Zabusky and M. D. Kruskal. Interaction of solitons in a collisionless plasma and the recurrence of initial states. *Phys. Rev. Lett.* 15: 240, 1965.

O. Zobay, E. V. Goldstein, and P. Meystre. Atom holography. *Phys. Rev. A* 60: 3999, 1999.

M. W. Zwierlein, C. A. Stan, C. H. Schunck, S. M. F. Raupach, S. Gupta, Z. Hadzibabic, and W. Ketterle. Observation of Bose-Einstein condensation of molecules. *Phys. Rev. Lett.* 91: 250401, 2003.